FRANGIBLE AMMUNITION

FRANGIBLE AMMUNITION

THE NEW WAVE IN FIREARMS AMMUNITION

JOHN F. MULLINS

PALADIN PRESS • BOULDER, COLORADO

Frangible Ammunition: The New Wave in Firearms Ammunition
by John F. Mullins

ISBN 1-58160-267-7
Printed in the United States of America

Published by Paladin Press, a division of
Paladin Enterprises
Gunbarrel Tech Center
7077 Winchester Circle
Boulder, Colorado 80301 USA.
+1.303.443.7250

Direct inquiries and/or orders to the above address.

Table of Contents

Preface

My interest in frangible ammunition comes from a very long time spent in firearms training. In 22 years in the U.S. Army, most of it spent in Special Forces, I trained indigenous personnel ranging from Montagnard tribesmen in Southeast Asia to Bedouins in Saudi Arabia. Another eight years working for the government were spent in the fields of VIP protection, high-risk installation security, and counternarcotics operations. Finally, I became that worst of all creatures, a free-lance consultant, still engaged in the work of preparing men for operations that would probably put them in harm's way.

What had been apparent to me in the early days became more and more clear as the years went on: there is simply no substitute for live-fire training. That wasn't so much of a problem back at Fort Bragg—and even less of one in Vietnam. We used live ammunition all the time in field exercises, counterambush drills, raid, and ambushes. In Vietnam we took our newly trained Strike Force troopers on patrol, hoping that the area we'd chosen had only lightly armed Vietcong units. Nothing like having someone shoot at you to ingrain the training you've received.

After Vietnam there was, of course, a new mission—several of them, actually. On the one hand, we still prepared for "the Big One," where we would go deep behind Soviet lines to organize, equip, and train indigenous forces and then lead them in guerrilla operations. But, frankly, no one ever expected that to happen.

What did happen was that terrorism reared its ugly head, and someone decided that the Special Forces were the logical nexus for the fight against this threat. We were thrown, willy-nilly, into operations that were just as different from the jungle warfare we were used to as a bar fight is from world-championship boxing.

The missions increasingly involved what could only be called "surgical" strikes. Instead of mounting an ambush in which the objective was to quickly annihilate everyone in the kill zone, we had to enter a building, take out the terrorists, and save the hostages and any innocent bystanders. Sometimes we didn't even want to take out the terrorists; we wanted to capture them and use the information gained to roll up the rest of the organization.

Of course we had to train for it. But how?

Our halting efforts, and the facilities we used (generally improvised), are described in Chapter 10 of this book. We became—and remain, especially given the added capabilities of Operational Detachment DELTA—a very competent force. But the facilities we built, or which were built for us, were lacking in a number of ways (also described later). The problem, of course, was that live ammunition has the nasty tendency to ricochet, splash back, and otherwise end up where it shouldn't. Facilities built for its use tend to be too small, too expensive, and too hard to maintain. Adapting other shooting facilities and trying to use them cause unacceptable danger to the trainees.

Instead of changing the training facilities, I thought, why not change the ammunition? Why not have a bullet that gives you the same effects as a conventional bullet but doesn't ricochet or splash back?

A bullet that is truly frangible.

I was neither the first to have this idea nor the last. I was probably the most underfunded, however, figuring that I could develop such a bullet in six months or so, and that it shouldn't take more than $20,000 or $30,000. Six years later, and more money than I care to talk about, I finally had a product. Only to find out that all sorts of bullets were being called frangible. There was, and still is, no simple standard that can be applied, no agreed-upon characteristics that make one bullet frangible, another not. Some bullets are being called frangible even though no one in his right mind would shoot them at close range against steel. Others may break up but shed copper splinters that can be more dangerous than the jackets from standard bullets.

This book is an attempt not only to define frangible, but also to inform the potential user of such rounds how to evaluate the ammunition, what it can and cannot be used for, and why you should consider adding frangible ammunition to your bag of tricks. With the correct round used appropriately, your training program can be vastly improved. And there are uses outside training that have only begun to be explored.

I hope that the information you gain here will be of help.

Introduction

Firearms science began when some 10th-century Chinese monks discovered gunpowder, it is said, quite by accident. Some say that they were looking for a recipe to preserve meat. One can only imagine the surprise of the monk when, upon pouring in the last little bit of saltpeter and having a stray cinder fall into the mixture, it blew up in his face. Of what possible use, he must have thought, would this strange concoction be?

It wasn't long before forward-thinking experimenters saw how the newly discovered force could be used to launch the bamboo tubes in which it could be contained, and, as a natural adjunct, projectiles from those same bamboo tubes. One supposes that some of those projectiles would have been smoothly worn river rocks, chosen because experimentation would have shown that they flew straighter than irregularly shaped objects.

Some time later lead was found to offer better ballistics than rock, and there the matter rested, with only incremental changes, until Capt. Claude Minié of the French army invented the bullet that still bears his name. During the same period self-contained brass cartridge cases were invented, vastly increasing the firepower of the individual. Then Alfred Nobel invented smokeless powder, which led to perfected high-power cartridges, which led to modern weapons designed to use those cartridges.

The bullets we fire today, regarded as state of the art, are not very much different from those being fired a hundred years ago, the changes since then being only incremental.

Until now, that is. The first viable revolutionary change to firearms science is upon us. This revolution has come about because of a pressing problem: the toxicity of lead. Lead, as a projectile, is being legislated out of existence and for a very good reason.

The new projectile, commonly called nontoxic frangible, has been made possible by advances in metal/plastic forming and offers a bullet that is as different from standard lead- and copper-clad lead bullets as the lead ball was from the rounded river stone fired by those early Chinese ballisticians.

Despite the rapid increase in acceptance of this ammunition by law enforcement and military agencies throughout the world, there has been no in-depth study of nontoxic frangible ammunition. The shooter has had to depend upon product literature, word of mouth, and his own experimentation when making a decision as to whether or not this type of ammunition is what's needed. It is this situation that this book is intended to remedy.

What is nontoxic frangible ammunition? Why not just say nontoxic and leave out the frangible if the elimination of lead is the only criteria? How is this going to change my shooting habits/training/use? Is nontoxic frangible truly the "round for all reasons?"

These, and many other questions, must be answered before the shooter can decide whether or not this type of ammunition is appropriate for the current need.

WHAT IS NONTOXIC FRANGIBLE AMMUNITION?

First, the nontoxic part of the equation. This projectile contains no lead or any other toxic heavy metals (as defined by national or international health and safety standards).

Shooting facilities already plagued with lead contamination will immediately recognize the value of this feature. Firing this round results in absolutely no hazardous by-products. The residue can be swept up and disposed of like the dirt on your floor (although this is not recommended, because the residue contains valuable materials that can easily be recycled for profit). No more Occupational Health and Safety Administration (OSHA) citations. No Environmental Protection Agency (EPA) hazardous materials findings. No raised levels of pollutants, other than a slightly increased level of carbon monoxide due to the combustion of powder (something that happens anyway). And for the shooter, no more worries about the buildup of a heavy metal in the bloodstream and bones, this accumulation being shown in numerous studies (see Chapter 3 for examples) to have severe and sometimes fatal health consequences.

Now for the frangible aspect. A frangible projectile is specifically designed to break into fragments upon impact with any surface as hard or harder than itself. Why is this a desirable characteristic? Any shooter who has spent time engaging metal-reactive targets can tell you why. The phenomenon known as "splashback," where pieces of the projectile come whizzing back to injure the shooter or bystanders, is so well known that it needs no explanation here.

A projectile with the proper degree of frangibility virtually eliminates this action. A properly designed frangible projectile should disintegrate into small particles, the largest of which contains no more than 5 percent of the original projectile weight, with the vast preponderance being much smaller. These tiny fragments lose velocity very quickly and virtually eliminate not only splashback but ricochets as well.

Every experienced trainer agrees that for true training transfer, practice must be done in conditions as close as possible to those encountered on duty. Tactical firearms training must be conducted with the weapon the officer will carry on duty and with ammunition that has exactly the same performance characteristics as the ammunition to be carried, in scenarios and training areas that mimic, as much as possible, the environments the officer will encounter.

To accomplish this task, and largely based on the experience of military trainers engaged in preparing specialized hostage rescue and antiterrorism units for high-risk missions, a number of avenues have been explored. Most of these involve the construction of "shoot houses" (often referred to as "kill houses" in military circles, a term rejected for civilian use because it was perceived as too bloodthirsty). To one degree or another, all possibilities have been less than satisfactory. The shoot house is either too expensive, too dangerous, too static, or too fraught with environmental problems (or a combination thereof). Chapter 7 will discuss in detail the problems encountered in these facilities.

But the biggest problem is the lack of access for the individual officer. Most shoot house practice is done by tactical teams because there is simply not enough training time or the facility is not available for the training of the "street cop." Yet it is this same street cop who routinely responds to domestic calls, silent burglar alarms, shots-fired scenarios, or other dangerous situations. It is the street cop who is first on the scene and is routinely the first to get killed. All too often by the time the tactical team arrives the situation is either resolved or stabilized, and it is the street cop who had to do an entry, a room search, or a movement under fire, something for which he is almost totally untrained under conventional training standards.

Enabling low-cost training for the street cop can only be beneficial to the officer, the agency, and the public, and there are few if any municipalities that do not have abundant abandoned buildings that have been condemned, seized for taxes, or forfeited. Many of these buildings, particularly commercial ones, are of concrete-block construction, which lends itself to use as a shoot house for officers using frangible ammunition. All that is needed is to block off any windows with mild-steel plate and use steel reactive targets inside. I recently converted such a building, comprising four large rooms, a cargo bay, and a rest room for the princely sum of $150. To date, more than 500 officers have been trained in this facility, all safely. No ventilation systems are needed. The bullets disintegrate upon impact with the steel targets and, if the target is missed, disintegrate upon impact with the concrete. Pieces of the concrete can be spalled off, it is true, but these pieces lose velocity almost immediately. The only maintenance requirement is to occasionally sweep up expended brass so that the next trainee will not lose footing on the expended rounds rolling around on the floor.

Another expedient in this low-cost shoot house is to erect take-down walls and room dividers of plastic and PVC, thus changing the layout and preventing the inevitable boredom that is the result of going through the same old scenario again and again.

For the officers we have trained—primarily individual officers (although we have also done teams)—the shoot house is a totally new experience. No longer are they standing in line on a range, punching holes in paper targets. Even the more progressive agencies, which try to provide at least a little tactical training on the standard range, do not normally allow their officers to move forward past a certain point, and so deprive the officers of the experience of tactical movement at close quarters. The intensity of the experience of our training course is illustrated by the sweat pouring from the officers in the dead of winter.

It is this type of training that nontoxic frangible (NTF) ammunition will enable. In the following chapters you will learn everything currently available about this type of revolutionary round, what it can and cannot be used for, the different types available on the market, how to evaluate the rounds, and how to discern your own need for such a round.

No bullet is a "round for all reasons." NTF is certainly not, despite claims to the contrary. Used as intended, however, it is a valuable tool for the trainer—for the administrator—and for the individual.

CHAPTER 1

Nontoxic Frangible Ammunition

A Round for All Reasons?

One of the problems with frangible ammunition lies in the definition. What, exactly, is it? The extent of this problem was shown a couple of years ago, when I attended a conference called by a major federal law enforcement agency (and potential customer). The purpose of the conference was to come to some sort of an agreement on standards for NTF ammunition, there being none in existence. It quickly became clear that frangible, on the part of both producers and consumers, was in the eye of the beholder. Well over half the three-day conference was taken up by arguments on exactly what should be called frangible and what should not, with the end result being that everyone agreed to disagree.

Does frangible simply mean that the projectile breaks up on impact? If so, a wide variety of ammunition could legitimately be called frangible. Technically, even copper-clad lead-core bullets (which from now on will be referred to in this text as "conventional" ammunition) could be called frangible because upon hitting heavy steel plates set perpendicular to the firer, the conventional round generally comes apart, the lead core and copper jacket separating and flying in different directions.

It probably makes more sense, then, to concentrate on the desired effect of frangible ammunition rather than on dictionary definitions. The desired effect is for the round to fragment completely upon impact. But how completely is completely? Some frangibles manufacturers say that since there is no such thing as complete fragmentation of any material that maintains its integrity while in flight, we should accept some lower standard for fragmentation. In this they are largely right. Modern "door-breaching" rounds used in shotguns do fragment completely, turning to dust practically upon leaving the barrel. This characteristic makes them useful for their intended mission, since with only a little standoff the round can be used to rip out locks and separate hinges from the door and frame, but will not harm someone standing only a few feet from the muzzle of the gun. Obviously, such rounds are useless in marksmanship training.

What lesser degree of fragmentation should we be willing to accept? One manufacturer states that "no remaining portion of the projectile should retain more than 5 percent of original projectile weight." Another argues that pieces of the projectile can weigh more than 5 percent but must not be able to penetrate certain substances set certain distances from the impact area. Still others argue that one must take into consideration not only fragment weight but also composition (a small, sharp fragment being capable of doing more damage than a larger, rounded one).

Again, intended effect comes to our rescue. We must ask what we want from nontoxic frangible ammunition. Is it the frangible characteristic itself, or are we more interested in the nontoxicity? And if it is frangibility, why is it necessary that the round be so constructed? In other words, what is the purpose of the round?

Perhaps this is the best spot to separate out the two characteristics: nontoxicity and frangibility. It is perfectly possible to have a round that is nontoxic but not frangible. It is also possible (though less likely) to have a round that is frangible but still toxic.

NONTOXICITY

Being nontoxic, under current definitions, means that the round should produce no residue that can be regarded as harmful to the human body. This includes every stage of the life cycle of the bullet: manufacture, storage, the act of firing the round, and cleanup.

That lead is a toxin we already know. Chapter 3 deals with this subject in more detail, but suffice it to say that we have more than enough evidence to indict lead for a whole host of health and environmental problems.

I will deal with lead toxicity during the firing process and subsequent cleanup because I feel that the reader is less interested in the problems of manufacture. There are two sources of lead in conventional ammunition. The first is in the primer itself. Most modern primers use lead styphenate as the explosive substance (some using a relative, lead azide). The lead does not disappear during the detonation process; instead it sprays out of the barrel in an airborne mist that lingers in the atmosphere for a while before coming to rest on any exposed surface.

The second source of lead is in the projectile itself. In a pure lead round (that is, not covered by copper or any other substance) the heat produced by the detonation of primer and powder vaporizes a small part of the lead base. Further lead is stripped off and atomized by the passage of the bullet down the barrel. Finally, when the bullet hits the trap (or other bullets, as in earthen berms on outdoor ranges) the heat caused by the energy transfer and the fragmentation process itself produces still more airborne lead.

Cladding the bullet in copper or other substances somewhat reduces the problem—particularly that of stripping off the lead in the barrel. Using a totally metal-encapsulated projectile reduces even more—primarily the effect of detonation on the base of the projectile. Neither cladding nor encapsulation eliminates the third source, impact downrange and the consequent spreading of airborne lead.

Using modern lead-free primers cuts down on airborne lead even more. Some ranges, heretofore unable to operate because of inefficient ventilating systems, have been able to reopen and maintain toxicity levels below that specified in OSHA and local regulations by using only totally metal-encapsulated projectiles loaded with lead-free primers.

But such solutions do not eliminate toxicity. There is still the problem of rounds downrange. Even if airborne lead levels are kept below toxic levels, there remains the problem of cleanup. Chapter 4 will deal with the problems of lead disposal in more detail, but it is sufficient to say that disposal is becoming more and more a problem virtually every day.

The only way to eliminate lead toxicity is to eliminate lead. But herein lie more than a few problems.

Lead, because of its density and low cost, is an almost ideal substance for a projectile. Lead projectiles fly farther, retain more of their energy during the entire trajectory of the bullet, and are capable of transmitting more foot-pounds of energy to the intended target during practically the entire process than are rounds of any other material (or at least any material that is affordable). The shotgunners among you will understand the frustration of handgunners and riflemen in trying to come up with an acceptable substitute for lead, having already had to go through the process of trying to find a replacement for lead shot.

So I will say this right at the outset: there is no ideal substitute for lead. Nor is there ever likely to be, despite any possible advance in metallurgy. That being true, it is also irrelevant. Lead is on its way out. The firearms practitioners I talk to regularly, people who do this stuff for a living, agree. In five years—10 at the most—lead bullets will be used only rarely. The toxicity problem is that great.

Those of us who have been searching for a substitute for lead have experimented with perhaps every metal known to man (at least those not radioactive or more expensive than gold). All these metals have problems.

FRANGIBILITY

Given all the following considerations for the types of ammunition I describe, we can come up with a usable definition of frangible. It is not merely a round that comes apart upon impact with a hard surface. It is, instead, a round that performs the tasks you want it to: most often, to engage targets at very close range and without the attendant danger of splashback and ricochet. The round must disintegrate upon impact into particles so small that they lose velocity almost immediately. Furthermore, the particles themselves must be of a shape that further decreases the possibility of injury; that is, rounded rather than sharp, with no jagged edges to catch and tear flesh. The particles must not produce excessive danger of splashback even when secondary impact zones are placed very close to the primary target (the "double-ricochet" effect).

With a definition that relies more on effect than on dictionary terms, we can start to evaluate the rounds we want to use for our firearms programs. Chapter 12 discusses in detail the methods available to evaluate the various rounds advertised as frangible. It is worthwhile to look at Chapter 2 to understand where this round came from.

Steel or Iron

People have been shooting steel bullets for more years than I've been alive. Every armor-piercing bullet contains a steel core of some type. So, why not just make the entire projectile from steel or iron?

The first problem is in barrel wear. Even if the steel or iron is softer than the material used in the barrel of the weapon, there will still be excessive wear produced by the passage of the bullet.

Then why not clad the steel core in a copper jacket? Fine, but then we still run into the problem of density. Ask any shotgunner if he gets the range from steel shot that he used to get from lead. Be prepared for an answer laced with profanity.

Worse, solid-steel bullets, even if clad with copper, simply could not be used on any indoor range. Not only would they seriously damage the backstop, but the ricochet problem would be horrendous. Steel doesn't deform the way lead does and doesn't expend all its energy at impact. Ricochets back to the 50-yard line would be common.

Solid Copper

Cast bullets made of copper have been tried as well. They eliminate the excessive barrel wear problem produced by steel, but that is about as far as it goes. Density is still not great, although to combat ranges solid copper maintains its energy fairly well. However, the indoor range problem remains. The solid-copper bullets will not wear out the backstop as quickly as steel but still will do so at an excessive rate. And the ricochet problem is still with us. Tests conducted a couple of years ago at the Federal Law Enforcement Training Center (FLETC) in Glynco, Georgia, were suspended when the rangemaster started finding spent solid-copper rounds back at the 25-yard line. One such round had flat spots indicating that it had ricocheted a total of seven times!

Zinc

This material comes closer to the density of lead. Cast-zinc bullets, either clad in copper or not, have been used in "gallery" shooting for a long time. If one is looking only for nontoxicity, zinc bullets loaded with lead-free primers can be an acceptable substitute.

Bismuth

This material, on first glance, seems to be an ideal substitute for lead. It is extremely dense and malleable, thus offering similar "mushrooming" effects upon impact with the target. It creates no excessive wear on the barrel. Unfortunately, bismuth is also, a fairly rare substance: thus it costs a lot more than lead (about $20 a pound in bulk, as compared to less than 50 cents a pound for lead).

Tungsten

Tungsten also has sufficient density, with an atomic weight of 183.85 as compared to that of lead at 207.20, but cannot be used by itself. Tungsten is one of the hardest metal substances known to man, much harder even than tempered steel. Rounds with cast- or machined-tungsten cores offer even greater penetration capability than do those with steel cores but will abrade a barrel within just a few rounds. No rangemaster in his right mind would allow a tungsten-core round to be fired at his facility.

The U.S. Army's recent "green bullet" initiative specifies a compressed tungsten-powder core, clad in a standard copper jacket. This alleviates the penetration and ricochet problem to be suffered with solid tungsten, but shares with bismuth another problem: cost. Tungsten is not as expensive as bismuth, but it is nonetheless considerably more expensive than either lead or copper. Perhaps the military, beset as it is with budget problems, can afford such a round. Most police agencies can't. Most individual shooters either can't or don't want to.

So we are left with few acceptable alternatives to lead, if the purpose is only to avoid toxicity. Zinc works, has an acceptable price, and is commonly available. But many shooters ask the question: If we have to change the material we shoot to avoid toxicity, why shouldn't we get some other benefits as well?

Why, indeed?

Thus it is quite common, when shopping for nontoxic ammunition, to find it to be frangible also. Modern frangible rounds use one or more of the substances discussed above but are manufactured by processes that eliminate the excessive ricochet, penetration, and splashback suffered from conventionally manufactured rounds.

Now we have answered a portion of one of our earlier question: Why frangible?

The remainder of the question is addressed in what we want frangible ammo to do, rather than what it is. Every rangemaster who has been in the business more than a few days can recite horror stories about the effects of splashback. Serious injuries, and even death, have occurred. This has become particularly acute in recent years with the increased use of steel-reactive targets. We like to shoot steel because of the immediate feedback it gives but are unable to shoot steel at the ranges in which lethal-force encounters normally occur. Seldom is there found a training facility that allows for the engagement of steel targets at ranges of less than 15 meters (16.25 yards). But lethal-force encounters normally occur at ranges of 7 to 21 feet. Is it realistic to keep training to the greater ranges? The obvious answer is no. But such training is necessary.

To see why, let's take a look at what happens when we fire our standard ammunition at steel.

Lead Bullets

Projectiles of solid lead are used far less frequently than

they used to be, largely due to the advent of semiautomatic weapons. Lead rounds tend to deform and to catch on loading ramps, increasing the chance of malfunction. But they are still used often enough in wheel guns (especially in the growing sport of Cowboy Action Shooting) to merit our investigation.

A solid-lead round fired against steel at an angle of 90 degrees has a tendency to flatten and deform, losing its energy almost immediately and most often falling to the ground at the base of the target. On the face of it, this seems a viable solution to problems of splashback and ricochet. However, in practice, the round often does not strike the steel plate at a perfect 90 degrees. At any other angle, even off the 90 degrees by slight increments, the round shows an increasing tendency to skip off the plate and ricochet. These angles are unavoidable if we want to engage the target from anything other than a fixed and relatively rigid firing position, or if we wish to engage multiple targets. The more off 90 degrees the angle, the greater the danger. Often, this danger is not posed by the initial ricochet itself but by secondary impact. The round skips off the target, hits one or more secondary objects, bouncing around until it finally loses its energy and comes to rest. All too often this is back toward the firing line.

Another problem, even when the round strikes at the desired 90 degrees, is the condition of the steel target itself. Repeated impacts against steel targets tend to "wallow" out the steel plate, making it concave rather than perfectly straight, and impacts by rounds of higher energy pockmark the steel. In both cases, such targets can "shoot back," directing the round back at the shooter or someone else on the firing line.

Copper Jackets

Lead-core ammunition covered with copper or another substance is even worse. When the round hits the steel target, even one in perfect condition and at an exact 90 degrees, you get another effect. The lead still flattens and deforms; however, during the process it literally "explodes" the copper jacket, sending pieces of it in a 360-degree spray that also causes secondary impact. The pieces themselves are often sharp and jagged, increasing the possibility of serious injury. And if the target's angle is at either more or less than 90 degrees, or if the target's steel is in less than perfect condition, you have the ideal situation for serious injury.

Therefore, rangemasters mandate shooting at 15 meters or more because it is likely that these pieces will lose their energy at this range—though this is not a given: injuries have been recorded at far greater ranges than 15 meters.

A properly designed frangible round eliminates this problem. And so we get to the crux of the matter: what is your intended purpose for the ammunition? If it is only to punch holes in paper targets or to shoot on ranges where the danger of splashback and ricochet is eliminated, then nontoxic rounds are probably all you need. If, however, you intend to perform tactical training at normal encounter ranges, frangible is almost a must.

CHAPTER 2

A Short History of "Unconventional" Ammunition

Frangible ammunition, despite its relatively recent appearance on the scene, is not new at all. The need for it has always been there. It is just that there were few workable solutions to that need.

World War II saw the first large-scale use of frangible ammunition, primarily as practice ammo for antiaircraft training. Practice for the gunners included firing at targets towed behind aircraft, the pilots of which were probably not too eager to be flying around in a blizzard of bullets fired by tyro gunners. Too, antiaircraft practice by its very nature takes up a huge amount of training space. Instead of being fired into a berm that is supposed to stop a bullet before it exits the training area, the rounds are fired at all angles into the air. Since the primary direct-fire antiaircraft weapon was of .50-caliber machine gun or something heavier, the rounds had tremendous range. The range safety zone for such practice had to cover hundreds of square miles.

So, lighter bullets were sought, something that could be fired accurately out to the ranges of engagement but would lose velocity and range quickly thereafter. As an ancillary benefit the rounds would, it was hoped, cause less damage, cutting down wear and tear on the intended target and soothing the fears of the pilots of the unintended targets.

The first substance to be tested was wood. Projectiles were turned out on lathes in appropriate calibers, loaded into the shells, and fired. Obviously, making such projectiles was extremely labor intensive, a problem when millions were being shot. Another problem was that the wood, even if it was relatively dense, did not offer sufficient range for anything but the most close-in shooting.

Later iterations turned to a zinc bullet. Zinc, with a specific gravity (SG) of 7.14 and an atomic weight (AW) of 65.39 is obviously far less dense than lead (SG 11.34, AW 207.20) and thus offers the quality of less maximum range than does the latter material. However, when bullets were made of zinc, a new phenomenon reared its ugly head. To build the appropriate chamber pressures to get the weapons of the time to function (particularly the .50-caliber Browning Machine Gun) the round had to be fired at much greater speed. At these increased speeds the round had even greater penetration than the copper-clad lead rounds.

The experimenters were running up against a practical application of Albert Einstein's famous equation: $E = mc^2$ (energy equals mass times velocity squared). Thus even a small increase in velocity has a far greater effect than does an increase in mass, the effect of the increase in velocity being squared, while the increase in mass is only added to the original.

This effect did not make the new rounds popular with the tow pilots. Particularly when the rounds, since they were cast, didn't come apart at all upon impact with the target. They had a far greater tendency to ricochet, with potentially disastrous results.

This problem continued to plague ballisticians. There was no doubt that there was a problem with standard rounds, even in the good old days when we didn't worry about lead pollution and its effects. Examination of earlier patents produces a litany of the woes to be expected when firing conventional ammunition. A particularly well-written example is found in a patent issued in 1989 to the inventors of a ceramic bullet.

> Firing ranges are typically used by persons practicing munitions firing, including military, law enforcement, sports, and recreational users. When such a firing range has been used heavily or for an extended period, using ordinary metallic bullets or projectiles, the area near the firing range can become dangerous because of the presence of large numbers of expended rounds embedded in the ground in the target area. These expended rounds can create danger by providing a hard surface from which new rounds can ricochet in an unpredictable and dangerous manner. The expended rounds can be removed by, for example, bulldozing, although at a large expense, particularly when the practice range is extensive, as in the case of a military aerial practice range.
>
> The danger from ricochets [is] not limited to ricochets caused by expended rounds. Projectiles such as bullets can ricochet from the ground or from a target even when the firing range is substantially free of expended rounds. For this reason, it is often desirable that practice munitions disintegrate upon striking the ground or upon striking a target.

A further difficulty with extensive use of ordinary metallic projectiles on a firing range occurs when a target is provided for practice purposes. In military or law enforcement practice, the targets often comprise such expendable or dummy objects such as vehicles, tanks, buildings, etc. Extensive use of such a target eventually results in the destruction of the target, requiring a replacement.

A continuing examination of patent documents reveals even more attempted resolutions to the problem—even though until relatively recently, the solution has been elusive. A few examples are cited here:

- Patent number 2,995,090, issued to Benjamin K. Daubenspeck on August 8, 1961, concerns a "gallery bullet" intended to be fired in the commercial arcades and amusement parks common during that era. Daubenspeck's bullet consisted of a powdered material in which the particles were substantially coated with a binder material of thermoplastic resins. The process he described was essentially a cold process in which the appropriate amount of metallic powder was mixed with the plastic and then formed. He claimed it was completely frangible, had adequate strength to resist breakup in handling and shooting in autoloading weapons, and would penetrate wood or other soft materials. No doubt this was true. However, in the patent itself the shortcomings of this process were revealed. In speaking of the .22 gallery rounds for which the process was designed, Daubenspeck says that the resulting bullet would weigh 15 to 16 grains, in comparison with the 28 grains of a lead bullet, just slightly over half the weight of a standard projectile. As a result, the rounds had to be loaded to substantially higher velocities, creating all sorts of problems, not the least of which was the danger of chamber pressures too high for the weapon.

- Patent 3,570,406, issued to Steyerberg and Kaseler on May 31, 1968, is another attempt to produce a frangible round, albeit one that comes apart almost immediately upon exiting the barrel. The purpose of such a round is questionable. Is it supposed to replace blanks?

  In any case, it is interesting because of later developments using a process similar to this, which will be described in future chapters. The Steyerberg/Kaseler bullet is composed of heavy metal granules, coated in a binder, and covered with a copper jacket. It is obviously not nontoxic, in that one of the metals envisioned for use is lead. The inventors claim that by encapsulating the lead they eliminate the problem of toxicity. One questions whether OSHA and the EPA would come to a similar conclusion.

- Patent 4,603,637, issued to James Snide and Edward Morrisey on August 5, 1986, describes a "Variable Density Frangible Projectile." The abstract reads: "A frangible projectile comprising a plurality of interfitting, filled polymeric portions is provided with controlled properties." What this means in plain English is a projectile made up of some 48 different components. The manufacturing difficulties in fabricating and assembling this projectile can only be imagined. Needless to say, we do not see the resulting ammunition on the market today.

- And finally, we have Patent 4,850,278, the aforementioned ceramic projectile, issued to Dinka, Jasa, Seegmiller, and Simmons on July 25, 1989. On the face of it, this seems a good solution. Modern ceramics can be made very dense and thus carry sufficient weight to operate the weapons without excessive velocity and chamber pressure. This projectile is made by sintering, essentially forming the projectile from the base material and a lubricant, then oven-curing the substance to melt out the lubricant. The resulting product is cross-linked and quite strong. The sintering process is commonly used to make self-lubricating bearings, the space left by the melted lubricant being filled under pressure with grease that is released as the bearing wears. Inasmuch as this is a mass-production type of process, the costs of manufacturing are reasonable.

However, when one considers what ceramics are, the problems are apparent. Ceramics are a glass-like material, quite hard, and very tough—likely, in fact, to be tougher and harder than the rifle barrel. Firing ceramic projectiles is likely to erode barrels so much as to ensure a lack of accuracy after a very few rounds. If the purpose is to make a smoothbore out of a rifled barrel, this may be your stuff. Otherwise. . . .

Shortly after this process was patented, what should be called the first viable frangible ammunition came on the market. But that is the subject for another chapter.

CHAPTER 3

Lead Toxicity—A Primer

One might ask (and I have heard many people do so) why we are so worried about lead. Lead is a natural substance and a quite common one. People live around lead mines and show no ill effects. Underground aquifers run all through the former mining town of Leadville, Colorado, and I have been told that the people show no ill effects from drinking the water.

All this may be true. It is also all irrelevant. OSHA, the EPA, and various other public health and safety organizations have taken the position that the ingestion of lead is a health hazard and have made rules that specify the permissible exposure levels (PELs). Unless someone can convince the entire regulatory apparatus otherwise, that is the way it is going to stay—with the possible exception of even more stringent rules.

That someone is going to convince these government agencies otherwise is unlikely, given the mass of data that has been accumulated on lead toxicity. Consider the following taken from OSHA Standard 1910.1025.

B. Effects of overexposure to lead.

(1) Short term (acute) overexposure. Lead is a potent, systemic poison that serves no known useful function once absorbed by your body. Taken in large enough doses, lead can kill you in a matter of days. A condition affecting the brain called acute encephalopathy may arise which develops quickly to seizures, coma, and death from cardiorespiratory arrest. A short-term dose of lead can lead to acute encephalopathy. Short-term occupational exposures of this magnitude are highly unusual but not impossible. Similar forms of encephalopathy may, however, arise from extended, chronic exposure to lower doses. There is no sharp dividing line between rapidly developing acute effects of lead and chronic effects which take longer to acquire. Lead adversely affects numerous body systems, and causes forms of health impairment and disease which arise after periods of exposure as short as days or as long as several years.

(2) Long-term (chronic) overexposure. Chronic overexposure to lead may result in severe damage to your blood-forming, nervous, urinary, and reproductive systems. Some common symptoms of chronic overexposure include loss of appetite, metallic taste in the mouth, anxiety, constipation, nausea, pallor, excessive tiredness, weakness, insomnia, headache, nervous irritability, muscle and joint pain or soreness, fine tremors, numbness, dizziness, hyperactivity, and colic. In lead colic there may be severe abdominal pain.

Damage to the central nervous system in general and the brain (encephalopathy) in particular is one of the most severe forms of lead poisoning. The most severe, often fatal, form of encephalopathy may be preceded by vomiting, feeling of dullness progressing to drowsiness and stupor, poor memory, restlessness, irritability, tremor, and convulsions. It may arise suddenly with the onset of seizures, followed by coma and death. There is a tendency for muscular weakness to develop at the same time. This weakness may progress to paralysis often observed as a characteristic "wrist drop" or "foot drop" and is a manifestation of a disease to the nervous system called peripheral neuropathy.

Chronic overexposure to lead also results in kidney disease with few, if any, symptoms appearing until extensive and most likely permanent kidney damage has occurred. Routine laboratory tests reveal the presence of this kidney disease only after about two-thirds of kidney function is lost. When overt symptoms of urinary dysfunction arise, it is often too late to correct or prevent worsening conditions, and progression to kidney dialysis or death is possible.

Chronic overexposure to lead impairs the reproductive system of both men and women. Overexposure to lead may result in decreased sex drive, impotence, and sterility in men. Lead can alter the structure of sperm cells, raising the risk of birth defects. There is evidence of miscarriage and stillbirth in women whose husbands were exposed to lead or who were exposed to lead themselves. Lead exposure also may result in decreased fertility and abnormal menstrual cycles in women. The course of pregnancy may be adversely affected by exposure to lead since lead crosses the placental barrier and poses risks to developing fetuses. Children born of parents, either

> one of whom were exposed to excess lead levels, are more likely to have birth defects, mental retardation, behavioral disorders, or die during the first year of childhood.
>
> Overexposure to lead also disrupts the blood-forming system resulting in decreased hemoglobin (the substance that carries oxygen to the cells) and ultimately anemia. Anemia is characterized by weakness, pallor, and fatigability as a result of decreased oxygen-carrying capacity in the blood.

Let's see. You will probably become impotent, but if you don't your children are probably going to be retarded. But you'll be too tired to worry about it, because you'll have anemia. Potential sexual partners are going to pass you up anyway, since wrist drop and foot drop are also signs of the central nervous system condition known as *tabes dorsalis*, most often brought on by third-stage syphilis. But the money you save on courtship will come in handy when you have to undergo dialysis, an expensive measure. You might not have to worry about it, though, because your brain will slowly die, your lungs will stop functioning, and your heart will stop beating.

What's not to like?

Of particular interest in this matter is the issue of the dangers posed to unborn children. One car battery manufacturer, in fact, used this information to ban all fertile women from lead-exposed jobs. For which the manufacturer was promptly sued by the United Auto Workers, which said that the policy was a form of sex discrimination that kept women from the higher paying jobs. That case went all the way to the Supreme Court.

How many police officers in your agency are fertile women? How many young soldiers, sailors, marines, and airmen also fit that category? Are we to ban these females—

most (if not all) of whom are in an age group that fits the maximum fertility cycle of females—from the firing range? The howls that would go up would dwarf those raised in the battery manufacturer-UAW lawsuit.

I discussed in Chapter 1 how the firing process spreads lead pollution. Now, how does the shooter (and to a much greater extent, instructors and rangemasters) ingest this substance? The most obvious method is by inhalation.

As stated in Chapter 1, the main explosive element in the primer is a compound called lead styphenate. The lead in this compound is not destroyed in the detonating process. Instead it sprays out into the air in microparticles, where it hangs for a measurable amount of time. Inevitably, you will breathe in a portion of this lead. But it doesn't stop there. More microparticles are made airborne by the action of the flame on the base of the bullet, the stripping off of lead in the lands and grooves of the barrel (if it is a solid-lead projectile), and finally from the cloud of lead produced in the impact zone.

Wearing a respirator is obviously not an option. Therefore, most indoor ranges spend a lot of money to design ventilation systems that take the airborne lead down-range, away from the shooter, where it is swept into an exhaust system that (it is hoped) is equipped with filters of sufficient size to trap the particles. Range designers are careful to plan firing points so that no dead zones or vortices are caused by booths, deflectors, or shields. And there the matter generally rests.

Although this kind of range system is better than nothing, problems remain. Some of the lead does not remain in the air, but sinks quickly to coat any exposed surface. Take a swab of exposed surfaces on even the best-designed range, and you will see it react to lead. This means that whenever you touch something, you are automatically contaminated. Not only that, but you are de facto contaminated yourself—

hands, face, uniform, boots, weapon. Given the unlikelihood of changing out of range clothes, taking a shower, and thoroughly cleaning every piece of equipment you had with you upon leaving the range, you are going to contaminate everything with which you come in contact.

And this brings us to the second way lead gets into your body—ingestion. Eat a sandwich without washing your hands and you are also eating lead. Smoke a cigarette and get the worst of both worlds, the carcinogens in the cigarette itself and the lead transmitted directly to the lungs, where it is quickly picked up by the bloodstream.

Perhaps even worse is the effect you can have on others. Go home after a range session. Walk in the house. You are dropping lead particles from your uniform and boots onto the floor. Strip your clothes off and throw them into the washing machine. You are contaminating all the other clothes in the machine. Comb your hair. You are dumping even more lead into the air and on the floor. Where do most toddlers spend the major part of their time? On the floor, of course.

Medical studies conducted by the National Institute for Occupational Safety and Health (NIOSH) have shown elevated levels of lead in the blood of children who lived in the same houses with fathers and mothers who frequented firing ranges, even when no other source of lead ingestion (e.g., lead-based paint) was present.

Many of us are cavalier about our own fate (the old adage, you've got to die of something, don't you?), but we are seldom so when it comes to the health of our loved ones.

Present OSHA standards call for a PEL of no more than 50 micrograms per cubic meter of air averaged over an eight-hour period. Any exposure above 60 micrograms for an eight-hour period is considered a *serious violation* by the U.S. Department of Labor.

Further, OSHA standards call for wearing a half-mask air-purifying respirator equipped with high-efficiency filters when the airborne concentration of lead is at or below 50 micrograms per cubic meter, and a full-facepiece air-purifying respirator with high-efficiency filters when concentrations are not in excess of 250 micrograms.

Before you say, "My range is modern and well equipped," consider the following report, taken at the behest of the U.S. Park Police at its 1992 indoor firing range. The study was conducted by NIOSH, and contains the following information:

> Air sampling revealed that for students using the range during training, the 8 hour TWA exposures ranged from 4.4 micrograms/cubic meter to 116.4 micrograms/cubic meter. Area samples ranged from 0.15 to 2291.1 micrograms/cubic meter.

In other words, if the students had been working in a factory where lead pollution was present, they would have been required to wear full-facepiece air-purifying respirators.

Many health and safety organizations state that even the 50 micrograms/cubic meter is way too high and recommend that this be halved. What would such a standard do to most indoor firing ranges?

Consider what happened when the old standards were passed. As one example, the Department of Defense conducted inspections of 1,800 indoor National Guard ranges during the period 1986–1988, resulting in the following: (a) 600 ranges deemed adequate, (b) 600 ranges shut down, and (c) 600 ranges requiring renovation to meet standards (at a total $90 million upgrade cost).

Not a problem you might say. We have an outdoor range.

Not so fast. Consider this. On a well-designed and tested indoor range, the air flow is conducted away from the shooter, minimizing exposure to airborne lead. There is no such ventilation on an outdoor range. With any luck, the prevailing wind will carry the fumes away. But the wind could also be carrying them right back into the lungs of the shooters or could be pushing the lead from one side to the other, with the lucky shooters upwind getting hardly any exposure and the unlucky ones on the other side getting a megadose. Or the day could be still, thus leaving the toxic mist to hang around the heads of everyone on line. In another NIOSH study, this time conducted at an outdoor range, the mean breathing zone lead concentrations for two range instructors were 67.1 and 211.1 micrograms/cubic meter, one instructor standing slightly upwind of the other and thus receiving a smaller dose, but both well over the OSHA standards of 50 micrograms/cubic meter. In another study, the breathing zone concentrations at an outdoor firing range were measured at between 460 to 510 micrograms/cubic meter. Four instructors at the range had blood-lead levels above 50 micrograms per deciliter of blood, far above the 40 micrograms considered the maximum allowable level. These instructors had to be reassigned.

Consider also that many outdoor ranges are semi-enclosed. Obviously, there must be a berm at one end to contain the bullets. Most often, the ranges will also be surrounded by fences or walls to ensure that no one enters the facility and to alleviate the noise pollution complained about by nearby neighbors. All these further constrict air flow.

And finally we get to another problem with outdoor ranges. The berm itself. The earth and sand berms normally used as impact zones in outdoor ranges very quickly get filled up with lead. In one recent study, some 41 percent of the mass in a berm consisted of spent projectiles.

Now another agency gets into the act. The EPA has taken the position that all those lead projectiles present a hazard to the environment. Rainwater leaching down through the berms, they say, picks up lead and carries it into the groundwater. Thus the lead must be removed, and some method found for containing contaminated water produced in the future. What this means in practice is that the berm must be removed and transported as hazardous waste to a hazardous waste disposal facility for processing. This goes not only for expended projectiles but for the earth itself, because it is now considered contaminated. And before the new berm goes in, the earth beneath it must be excavated, an impermeable layer put down, and a pumping system installed to get rid of the water that will accumulate there. This water must then be filtered to remove all trace of lead particles before it can be released. Only then can the berm be replaced and firing continued. Few ranges are willing to take these steps.

One can argue, and many have, that the risk of groundwater contamination from this source is vastly overstated. That may be true. It is also, as we stated at the beginning of this chapter, irrelevant. Try bucking the EPA with such an argument. After you've paid the ever-increasing fines and sanctions it will impose upon you, your argument will seem rather useless.

And lead isn't the only problem.

CHAPTER 4

Other Firing Range Problems

Lead pollution is not by any means the only problem besetting firing ranges, both indoor and outdoor. The problem starts with the way they are built.

Most firing ranges are modeled after military marksmanship ranges, the military having been in the forefront of firearms training from the beginning (for obvious reasons). Is the military marksmanship range model readily adaptable to civilian/police use? The answer is, all too often, no.

One of the problems lies in the ranges' locations. The military model envisioned almost unlimited space for its ranges, with "range fans" of hundreds of square miles not uncommon. Look at the engineer maps of any active army base and you will see huge areas with restricted access; the military always expects that some rounds will escape the confines of the range proper and make their long way downrange.

Civilian/police ranges seldom have this luxury since most of them are located in urban areas. Try to convince the city fathers that you need a range fan extending a couple of miles beyond the berm of your local range, where no one can enter and certainly no development can take place.

And even the military, increasingly affected by encroachment on its old training areas, is searching for a new paradigm. Quite simply, there is never enough space. With the base closing taking place every few years, increasing numbers of troops have been concentrated onto the bases that remain. Even with the military's current downsizing, the problem of finding training areas for all those troops remains.

On the typical outdoor pistol range, the berm or backstop is a 20- to 30-foot earthen mound, in front of which are fixed target positions. In most cases there are concrete-surfaced firing lines 3, 7, 15, 25, and 50 yards from the target line (though many newer ranges that were not designed for Police Pistol Competition [PPC] type of competitions have eliminated the 50-yard line). The various firing lines have positions marked to correspond with numbers on the targets. At the 25- and 50-yard lines there are usually wooden 2 x 6-inch posts that are supposed to serve as barricades, either permanently fixed or inserted in mountings that allow for removal and replacement.

The target systems are either of wooden or metal frames set in the ground to electrically or pneumatically activated "turning target" systems. Occasionally, a range is also equipped with some type of "running man" or moving target system, generally mounted just behind the normal target line. And some ranges have been augmented with independent target systems designed for use with picture-type targets, facilitating decisional-type shooting.

The standard range was designed and works well for basic marksmanship training, but it is woefully inadequate for all but the most rudimentary forms of practical and tactical training. As an example, officers are taught the hip-height, tucked-pistol position for close-contact or "interview stance" firing. Just a slight elevation of the muzzle can cause rounds to hit high on the berm, striking at such an angle that they skip over the top (causing no end of consternation downrange). To prevent this, most ranges of this type have been forced to install expensive wooden "brows" on top of the berms.

Few of these ranges allow for the engagement of multiple targets, particularly at normal engagement ranges. On standard ranges, muzzle offset is allowable at no more than 30 degrees from center; otherwise, rounds will have a tendency to exit the range from the sides. At 25 yards' distance, targets

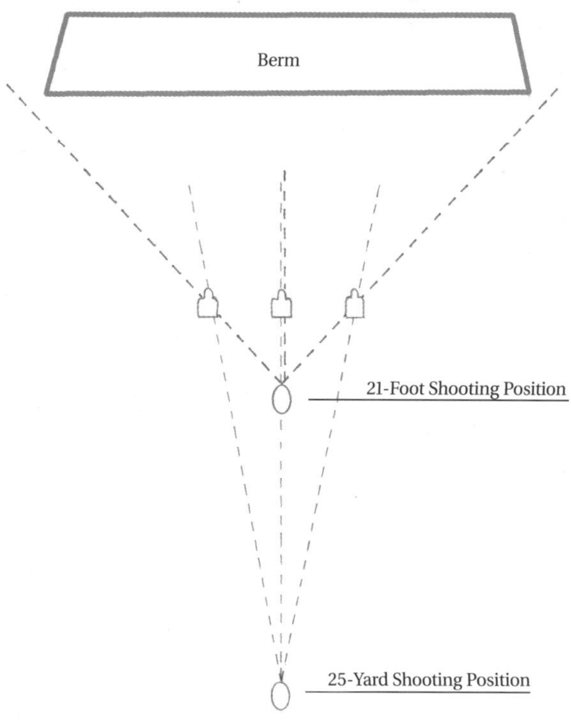

can be placed in multiples that stay within the 30 degrees. At 7 to 21 feet, this is practically impossible to do (see previous illustration). For all practical purposes, this means that multiple-target training must be conducted no closer than at the 7-yard line, and at this line the student can only engage two adjoining targets. To engage targets farther apart, which should be done because bad guys aren't often cooperative enough to stand close together, the distances from the target must be increased. Although this sounds like a realistic solution, the officer is still not being trained to move the muzzle more than 30 degrees. If an instructor wants to produce training that allows for more than 30-degree movement, an expensive angled apron must be added to the range.

Even with a brow and apron added to the range, there is no guarantee that the rounds will be contained. Berms are rarely 100-percent earth. They typically consist of a rock core overlaid with tires, a layer of dirt, or sand covering the latter. Extensive use and poor maintenance may expose the core, but this is not the most common problem. The expended projectiles themselves, if not periodically mined, form a semisolid layer of lead and copper just beneath the dirt surface. As new rounds are fired, they strike the expended rounds and either start splashing back at the shooter or skipping over the berm. The wooden brow can stop the rounds from skipping over the berm, but many brows develop gaps or are worn through from repeated projectile impact or weathering, with aprons suffering the same problems. As for an expended round found downrange of the berm, particularly if it is in a civilian area, ask any police agency that has been through that ordeal.

It is not only the berm that can cause rounds to leave the range or splash back at the shooter. Metal target frames and mechanism protectors also cause projectiles to ricochet, skip, or splash back. In one incident a handgun projectile struck a metal plate, passed over the brow and berm, and entered a trailer 400 yards past the berm. (The trailer was being used as an office for a construction company.) Luckily no one was hurt, but the construction company obtained a court order forcing the range to close.

Added to the problem is the fact that many ranges, particularly outdoor ones, are also being used for rifle training. With the increase of perpetrators wearing body armor, more and more law enforcement agencies are adding rifles to the weapons available to the street officer. But many agencies considering this addition will not determine until it is too late whether their ranges are capable of accommodating the training necessary for these weapons. The majority of existing law enforcement ranges were originally designed for use with weapons with a maximum range of 1,700 to 2,000 yards (representing the maximum ranges of commonly used law enforcement pistol ammunition). The .223 Remington, the weapon of choice for most police agencies arming themselves with rifles/carbines, has a maximum range of 4,405 yards. And since most agencies do not have control of all property within the maximum range of their handgun rounds, not to mention the rifle ammunition, one can readily see the problem. Too, the wooden brow designed to stop handgun rounds is easily shot through by rifle rounds, particularly the ball rounds most commonly used in training.

The upshot is that even though the ranges are used to improve the skills of the officers who protect them, citizens don't care about this when they find an expended round on their property. As well, encroachment has restricted the use of (or has closed) many law enforcement ranges and will affect more in the future.

Environmental concerns, as we have already seen, inevitably affect the outdoor range. Lead must periodically be mined from the butts/berm and disposed of as hazardous waste. Recent rulings by the EPA have virtually eliminated the old practice of selling the lead to a scrap-metal dealer.

Now there must be a documented chain of custody in which the lead is traced from its removal point to the final destination. Scrap-metal dealers are seldom willing to go through the administrative hassle, and fewer still are willing to place themselves in the way of a potential liability that can haunt not only them but their descendants.

OSHA gets into the act, too, and mandates certain safety requirements when removing the lead. Masks with HEPA filters are only a part of it. Observe the removal of lead in an OSHA-approved manner and realize that asbestos remediation, by comparison, is a fairly mundane task. Not only do the actual workers have to follow extremely strict standards, but the site itself must be quarantined, methods put in place to eliminate the hazard of dust from the removal effort wafting into areas where people might be exposed, and constant testing be done.

Little wonder, then, that the recent effort by New York City to remediate the outdoor range complex in Rodman's Neck, in the Bronx, was estimated to have cost more than $100 million.

Oh, and those expensive wooden brows and aprons? They are also now contaminated with the lead lodged in them. Dispose of them by burning? Get caught and expect a sanction that will bankrupt your agency. Imagine the effort that will inevitably take place when such brows as those at the St. Louis Police Department range must be disposed of. They were installed when a number of expended rounds were found on the rooftops of businesses across the interstate highway from the range. These brows are of stout oak, each portion of which is 12 x 12 inches by some 20 feet, and extend the full length of the range, to a height of some 30 feet. Close inspection of these brows shows that they are riddled with lead bullets.

Larger or more affluent agencies have therefore built indoor facilities as a solution. Because they offer a controlled environment, including lights and temperature, the ranges are well suited for teaching basic marksmanship.

Unfortunately, indoor ranges are even more poorly suited than outdoor facilities for advanced law enforcement training. Most backstops do not allow for angle shooting, so multiple-target engagements and moving target drills are impossible. Even fewer backstops are rated for rifle calibers, and certain brands of 12-gauge slugs may dent the armor plate. Hip-style close-combat shooting done from the normal firing line at close-in targets can cause damage to overhead lights and baffles, and moving downrange for close-in shooting near the target line is not possible because of the danger of backsplash.

And forget about moving and shooting, taking advantage of available cover, and so on. Rounds fired during such exercises may or may not hit the bullet trap, many of them instead hitting the concrete block typical of sidewall construction on indoor ranges, where they will in a worst-case scenario, ricochet and injure the shooter or bystander, or in the best-case scenario, damage the blocks so that they will soon have to be replaced. Also at great expense.

Typical firearms ranges, in general, make realistic tactical training an impossibility unless expensive modifications are performed. Obviously, 360-degree shooting is out. Even ranges that have aprons running back the entire length of the range allow only about a 180-degree shooting zone. Chapter 7 deals with past attempts at tactical ranges in more detail, but it's enough to say that changing a standard range to something where a tactical team can perform live-fire practice is an expensive proposition. Further, due to the lack of portability in these solutions, the regular range becomes severely constricted for standard use. Few ranges have the luxury of enough space to accommodate both types of training.

Now let's consider one of the biggest problems facing the firearms instructor today: cost. Money. Filthy lucre. And the absolute lack thereof.

As can be seen from the earlier discussion, building a new range for all but the most flush of agencies is going to be impossible. First an area must be found that, essentially, no one else wants. With the ever-increasing cost of real estate and continuing urban sprawl, finding a place like this within easy commuting distance of the city is quite a task.

Then the site must be prepared. Every tree, rock, piece of junk, or anything else that might cause ricochets or backsplash must be removed. The ground must be leveled and graded, removing any obstacle that might cause an officer to stumble while carrying a loaded weapon. Of course, drainage must be ensured; little training is going to take place in a swamp.

To meet EPA requirements, the berm area must be excavated and equipped with a catch basin lined with impermeable plastic or rubber. On top of this goes some sort of rough media through which water can freely flow, and which allows its evacuation through a pumping system. The water must then flow through a filter that removes all traces of lead. Only then can the berm be installed. The berm itself must be scientifically designed to catch all rounds that hit it, without danger of ricochet. Aprons are then built, also to specification, but do not have to meet the same drainage standard as the berm, unless they are intended to be fired at directly (as would be done in tactical training). Brows are then installed above the berm and aprons, or the range is equipped with a series of wooden baffles through which no round from the firing line can escape, even when shot at acute angles (as sometimes happens with accidental discharges).

Finally, the entire range is graveled, has concrete firing lines installed; covered firing line positions are constructed; and a range house, storage, and maintenance shed are built.

And I haven't even gotten to the installation of targets and target systems yet.

All this is done out of a budget that is already constricted by a hundred competing requirements.

It doesn't stop there. As mentioned earlier, the lead in the berm must be periodically mined, using hazardous material (hazmat) disposal procedures. The lead and contaminated earth must be disposed of, but even earlier than that the filters in the drainage system must be removed and disposed of, also using hazmat procedures. The wooden brows and baffles must be continually inspected and, if worn, replaced. Target systems are inspected and, all too often, replaced. Even pistol bullets do tremendous damage to target holders, turning and raising mechanisms, and any sort of other equipment downrange. (Is there a range officer out there who has not seen a shooter deliberately target things he is not supposed to be shooting?)

Maintenance costs can often run more than the cost of the range itself. And those costs just keep on coming.

Indoor ranges? Take the costs for an outdoor range and multiply them a few times. If you wish to shoot a rifle or shotgun, as most agencies do these days, the backstops must be rated for rifle and shotgun—at considerably more cost. Space is necessarily even more constricted. If you wish to fire at more than what's just in front of you, the side walls must also be equipped with either bullet traps or one of the newer antiricochet systems currently marketed. Costs are, of course, extremely high.

You don't have to worry about groundwater now, but you must still get rid of the lead that accumulates in the bullet traps. No longer can you depend upon the good-ol'-boy who came by periodically and actually paid you for the privilege of removing the spent lead.

Of course, you have to have installed an extremely expensive ventilation and filtration system. The system will be tested to make sure that there is positive airflow past the shooter and that there are no dead spots, vortices, or up- or downdrafts. The air that gets pushed past the shooter will then have to go through a HEPA filter of sufficient size to provide positive scrubbing of all the air that passes through it, and the filter must be replaced regularly. (Of course, the filter itself is now hazmat.)

Climate control in such a range is a huge problem. All HVAC (heating, ventilating, air-conditioning) systems depend for efficiency on the recirculation of air that has already been heated or cooled. So, the intakes for the system have to be past the filters at some point; otherwise contaminated air is pushed back into the shooting/administrative zone. Only the most efficient of filtration systems allows this. Remember, to ensure no leakage of lead contamination to the outside, the inside of the range must be kept in a slight underpressure situation. Thus when a door is open, the air from the outside comes into the range, rather than the reverse. HVAC systems generally work the opposite way, to keep outside untreated air on the outside. Often the design of the filtration and HVAC systems costs more than the range itself.

The range must be cleaned periodically because even with the best of ventilation and filtration systems, lead is still going to settle on any exposed surface. Workers will have to wear full-containment suits with breathing systems that can block any lead contamination. The most efficient way of cleaning a range of lead is good old soap and water. Now what do we do with the contaminated water? You guessed it.

By the way, even if you have the best designed and maintained range in the world, you still must take blood samples from range personnel on a regular basis to ensure that their blood-lead levels are not above what is safe. If by chance any levels are, affected personnel must immediately be reassigned. And if they have health problems in the future, even if those problems have nothing to do with lead, expect workmen's compensation claims, at the minimum. At the maximum? Well, look at the ads for personal injury lawyers being aired on television.

And these aren't the only problems.

CHAPTER 5

Training Philosophy

The purpose of law enforcement training is this: to prepare the officer to handle situations likely to be encountered in the performance of duty. This is often expressed with adages such as "As ye train, so shall ye fight."

Obviously, then, all firefights occur in broad daylight; the officer is well rested, not in a state of fear; the target is not moving and certainly is not shooting back; the weather is fine; adrenaline is not pumping through the bloodstream like a river of fire; and the hands are not so sweaty that it feels like the gun could slip right through them.

No?

Then why do we train that way?

In too many cases, the final critique of firearms training comes in a civil courtroom. In fact, firearms are second only to vehicles in causes for civil damages for law enforcement agencies. Two cases warrant examination here: one because it is a landmark, and the other because it addresses "modern" training practices.

First, *Popow v. The City of Margate (N.J.).* In this case, two Margate police officers were involved in a foot chase with an armed suspect being sought on kidnapping charges. The chase was at night in a residential area, and the suspect and the officers exchanged shots. Mr. Popow, a citizen, was attracted by the noise and stepped out on his porch, where he was killed by a stray bullet fired by one of the officers. The widow Popow sued under the provision of 42 USC 1983 (civil rights). The case was heard in U.S. District Court in New Jersey, and the court found the city of Margate liable. The city and the officer involved moved for a summary judgment (dismissal), arguing that there was no claim under 42 USC 1983. The city argued that there was no evidence of unconstitutional official policy ("Under the color of any law . . . ."). The judge denied the motion and, in addressing the city's claim, indicated that grossly inadequate training and supervision could, in fact, constitute an official policy on the part of the city.

The judge referred to the following circumstances as grounds on which the court could conclude that training and supervision were grossly inadequate:

1. The original firearms instruction for the officer who fired the fatal shot took place 10 years prior to the incident.
2. The only continuing training on shooting took place biannually.
3. The city had not trained the officer to shoot at moving targets.
4. The officer had not received training on how to shoot at night.
5. There was no training or guidance on shooting in a residential neighborhood (decisional training).

6. The city used no films or simulation to teach officers how law, regulation, and policy were to apply in practice (decisional training).
7. There was inadequate communication of policy.
8. Disciplinary procedures were inadequate.

A more recent case, *Zuchel v. City and County of Denver*, focused again on training, but this time the court was even more critical. Zuchel, with his hands in the air, was shot four times by a Denver officer. The jury award of $330,000 and five years of attorney's fees was upheld upon appeal. During the trial, it was disclosed that the officer had received his only decisional shooting training through a video projection system. This system projected various situations on a screen, with the trainees reacting to situations with firearms modified to emit a laser beam when a primed case was struck. The point of impact of the beam was recorded, and when the scenario was played out there was a critique of the trainee's performance. The basis for the court decision was the Denver Police Department's failure to do live-fire decisional training. The court found that this was deliberate indifference on the part of policymakers.

Imagine a headline similar to this one, taken from the *Daily Oklahoman* on July 22, 1997, appearing in your hometown paper.

Court Upholds Widow's Lawsuit against Police Officer's Training Questioned in Slaying

Mr. Allen was depressed over the death of a relative a few years prior to February 20, 1994. On that date he had a fight with his wife and children, and after the fight left the house with several firearms and ammunition. He later parked in front of his sister's house, and a 911 caller reported to police that Mr. Allen was contemplating suicide. Four officers,

> including one lieutenant, responded and found Allen in his car holding a gun. Lt. Smith told Allen to drop the gun and then the other three officers attempted unsuccessfully to seize the firearm. Allen then turned the firearm toward the officers who fired twelve rounds, four of which struck Allen, killing him. Allen's widow sued, claiming that the officers used excessive force and that the City of Muskogee had failed to properly train them.

A U.S. District Court dismissed the case, but the U.S. 10th Circuit Court of Appeals reversed that decision. In the ruling the court judges wrote, "A reasonable jury could conclude on the basis of some of the testimony presented that the officers' actions were reckless and precipitated the need to use deadly force." Some of that testimony was presented by Dr. George Kirkham, a former police officer and a police training expert.

Dr. Kirkham stated that the officers should have taken cover and tried to communicate with Allen. A portion of his testimony said that "if the State of Oklahoma through CLEET [Council on Law Enforcement Education and Training, the state agency in charge of officer certification] trains their officers to not be mindful of cover, to be aggressive with mentally ill people, to do all these things that are totally off the board, then the state is completely out of sync with the rest of the country in the police profession." Mrs. Allen is suing for an undisclosed amount.

I suspect that many of you have already had such headlines in your area. Or worse. Why is that?

"As ye train, so shall ye fight."

But we do not have to concentrate on legal liabilities to see the inadequacy of many if not most training programs. When under duress, we revert to those actions we were drilled in. Back in the days when most police officers were carrying revolvers, there were incidents when, after a lethal force encounter in which the officer came out in second place, he was found with six expended shells in his pocket and scattered live rounds around his body. Why was that? Because on the range he was taught to empty his weapon at the target, shuck the expended rounds from the cylinder into his free hand, place the casings into his pocket (the rangemaster not being happy when his beautiful facility was sullied by loose brass), and only then reload. The assailant, seldom worrying about expended brass, simply waited until the officer was out of ammunition and then came over and shot him in the head.

Enough of those incidents and we got away from that, but there are other practices being taught that are just as deadly. Officers stand out in the open and return fire, hoping their superior accuracy will win the gunfight. Why is that? Because they've been trained that way, although that wasn't the initial purpose of marksmanship training. When you're on the firing line popping rounds at the target, you are not being taught to be situationally aware, to look for cover, to move to the right or left to reduce your exposure, and to "line up" your assailants so that a two-assailant attack becomes a one-assailant-at-a-time attack. You are instead concentrating on your stance, your wrist lock, your sight picture, your trigger squeeze. Just as you've done all that time on the range.

I'm often asked why it is that police officers, with all the training they've received, consistently lose gun battles with felons who may have never fired a shot before?

Think of the constraints on each police officer and then those on the felon. The officer must first make the decision to apply lethal force. He's been drilled a thousand times about this decision. And while all the elements contributing to the decision are running through his mind, the felon has already made his own. He knows he's going to shoot. The officer doesn't. Because of the critical time factor, all too often, the felon gets the first round off, and the first round often decides the gun battle.

Even when the officer has made the decision to apply lethal force, he still faces constraints. He must make sure of his target and beyond. He must ask himself the following questions: Are there innocents who may come into the path of my bullets? If I overpenetrate, will this round hit Granny down the street? If I miss, will the ricochet skip around until it goes through someone's car window?

The felon, on the other hand, really doesn't give a damn about whom else he might hit. He is a felon, after all.

Make every shot count, the officer has been told. The felon knows that he only has to make one shot count.

And, finally, when we get to the shooting itself the officer goes back to his marksmanship training. Sight picture, good stance, trigger squeeze, follow-through. Even when the encounter, as most are, is within 7 to 21 feet.

The felon, since he has not been taught all this, moves to instinctive shooting. Thrust the weapon out as you would your fist, squeeze the trigger in a convulsive grip, blast away. He has instinctively gone to the technique taught by the late, great Col. Rex Applegate to OSS operatives. It is an extremely fast method of shooting and, at close range, lethally effective.

Some "pundits" attribute the shooting skills of felons to a number of extraneous factors, one even going so far as to say that video games are teaching our kids to kill. Those of us who struggle with getting true training transfer from even paintball guns and marking-pellet-modified weapons recognize that for the nonsense it is. First, the idea that manipulating a computer mouse is in any way analogous to firing a weapon shows arrant ignorance. Second, the phenomenon of winning a gun battle by using instinctive shooting has been going on since long before computers were even invented.

Would that such silly analyses were true. We could get rid of a lot of very expensive training, sit each officer down at a computer console, and have him or her play off-the-shelf first-person shooter games. No ranges, no ammunition, no worry about lead, no instructors, no inclement weather schedules. It would be very nice.

It isn't going to happen.

Now, we cannot get rid of the first factors influencing the officer involved in a lethal force encounter. He still has to make the decision that lethal force is appropriate. He still has to worry about targets other than the intended one. But he should not be saddled with the inability to use his weapon any other way than what was taught on the marksmanship range. Obviously, initial marksmanship training is essential, but it must go further than that.

One of the few positive things that came out of the Columbine High School tragedy in Littleton, Colorado, is that law enforcement has belatedly begun to recognize that there are situations in which you simply cannot sit back and wait for the SWAT team. What had first-responders been trained to do prior to this realization? Contain the situation, secure the perimeter, wait for help. Some of them are still being taught the same.

At our training facility, we get a heavy contingent of local sheriff's deputies and small-town police officers. Being in a fairly underpopulated area, there aren't the resources available to form special reaction (or whatever the buzzword is these days) teams. Our only team of that type comes from the state police. Response time is, of course, fairly long.

I pose a scenario to these students. You've responded to a shots-fired situation in a local hospital. What do you do? When I get the inevitable answer of containment, perimeter security, and, sometimes, with the brighter ones, situation assessment and the gathering of information, I then pose the second part of the scenario. You know that your wife and newborn baby are inside. You think you hear your wife's scream. Now what do you do?

Now realize that it is *somebody's* wife, *somebody's* child. And you're going to tell them you can't go in, that it doesn't matter that the person's loved ones are in mortal danger?

And then you'll wonder why the police get criticized for inaction?

I know all the arguments. It doesn't do any good for you to be added to the list of casualties. You may make the situation worse than it already is. You don't know who the felon is and might shoot the wrong person. We could go on for the rest of the book listing the excuses I've heard. The fact is, all these excuses stem from one root cause: lack of training.

The first officer responding in such situations is almost always the average street cop/patrol officer—and it's that cop who has to make the split-second decisions that may mean the difference between life and death. Even by the time that close-by backup has arrived, the situation has often played out. And it almost certainly has when the special reaction team arrives. Its response time, in the best of situations, is measured in hours rather than the seconds it takes to make the difference. And yet the street cop/patrol officer gets the least training (with the possible exception of administrative personnel) of anyone on the force.

Columbine, as mentioned earlier, is changing training slightly. Such situations have produced a new buzzword, "active shooters." The term, as defined in a recent Illinois Tactical Officer Association newsletter means "one or more subjects who participate in a random or systematic shooting spree, demonstrating their intent to continuously harm others. Their overriding objective appears to be that of mass murder, rather than other criminal conduct such as robbery, hostage taking, etc."

Police officers in many agencies are now being trained to deal with active shooter situations, the difference between present and past training being the recognition that one cannot wait for the tactical team to arrive. The first officers on the scene must respond, and respond quickly, before there is even more loss of life. The officers are being trained to form ad hoc entry teams that are able to clear rooms on their own and engage the active shooter (or at least bottle him up so that no more harm is done to innocents) on their own. Such training has to be standardized: you don't want officers from different jurisdictions arriving on the scene, forming ad-hoc teams, and finding out that their procedures are completely different.

One can only applaud this latest development in training. It's a shame that it took a tragedy to implement it.

The only problem with such an approach is that definitions tend to take on a life of their own. Are we going to use these tactics only when "their only purpose seems to be that of mass murder, rather than . . . ."? Under such a definition, the North Hollywood shootout, in which two bank robbers armed with automatic weapons and covered with soft body armor took on the police forces of several jurisdictions, would not have qualified as an active-shooter situation. Although many pundits have said the robbers' motive was to kill cops, that wasn't so. Had it been their intention to kill cops, there would have been many more dead cops. Any infantryman could tell you what actually happened. The robbers used suppressive fire to keep any cop's head down until they could escape.

However, a bullet coming at you doesn't really care if its purpose is suppressive fire or if it has been shot with the intention of its lodging firmly between your eyes. The effect is the same. Thus we must train police first responders to take into account all situations where shots are, or are likely to be, fired.

But we don't. And why is that?

CHAPTER 6

Current Training Methods

Initial firearms training consists of 40 hours of weapons handling, maintenance, and basic marksmanship. During this 40-hour course, rookies are generally exposed to the handgun and the shotgun. In the past, this training was considered adequate because most recruits had military backgrounds and training, and thus the basic concepts of marksmanship were familiar to them. Today, typical recruits have never touched a firearm, and their only exposure to lethal weapons is what they have seen on TV and in the movies.

After the recruit leaves the academy, firearms training tends to become sporadic, at best. Surveys taken of law enforcement firearms instructors indicate that few are providing as much of, or the type of, training they would like to. In all too many cases there is no continuing training at all, firearms being used only on the periodic qualifications required by state and local mandates. Even in cases where there is continuing firearms training, it is all too often repetitive marksmanship practice.

There are a multitude of reasons given for this deplorable status, most of which fall into categories summarized by the acronym TIME:

T — Time itself, or the lack thereof
I — Instructors, and why their usual qualifications are not up to the job
M — Money, the lack of which forestalls even the best of intentions
E — Equipment, and the inadequacy thereof

Rarely is the problem rooted in just one area but is instead a combination of the factors. Let's take a closer look at each, and see how it affects firearms training.

TIME

In the past, it was not at all unusual for firearms training to be conducted during the officers' off-duty hours, the argument being that there are simply not enough hours in the day for the officer to patrol, fill out paperwork, show up for court appearances, take annual leave, account for sick time—all the time-burning events that plague staffing planners—and requalify with firearms as well. However, several court cases have ruled that even public employees have to be compensated for such training time with overtime pay. With salary budgets shrinking, this practice has pretty well disappeared. Now the time for firearms training must come out of regular duty hours. This creates an even more severe time problem, and one that may differ with the size of the agency.

For larger agencies, the time issue often combines with other problems to present a roadblock to training. The average law enforcement agency already suffers a severe shortage of officers. Generally, the number of sworn officers is well below authorized strength, and ever larger numbers of officers are being assigned to "social" duties. Administrators and supervisors tasked with conducting traditional law enforcement tasks find their ranks severely thinned. Court, vacations, holidays, and sick leave further deplete resources to the point that they start to view any activity that takes officers from their assignments as something to fight. This includes training.

For smaller agencies the plight is often even worse. If an officer is off the street for any reason, it means that a small town or a section of a county goes unprotected. If an investigator is otherwise assigned, cases go ignored. Although many chiefs, sheriffs, top administrators, and supervisors support aggressive and active training programs, they simply cannot allocate the necessary time and manpower for training.

INSTRUCTORS

Agencies often forget that the background and training of instructors have a direct influence on the training they give to others. Many agencies select firearms instructors based on scores from the firing range, and the result is that the best shot becomes the instructor. Once selected, the instructor is usually sent to a basic firearms instructors course that satisfies the state law enforcement training board. The instructor is then expected to combine personal shooting ability with the methods of instruction just learned in the basic course to produce effective firearms training. Although there seems to be logic in this, it is all too often an error.

It is true that qualification scores are a test of marksmanship and that an officer who consistently scores high demonstrates an understanding of, and the ability to apply, the basics of shooting. In fact, the ability to consistently apply these basics is almost always the result of practice, and most shooters who practice a lot tend to be firearms competitors. At first glance, having an instructor who is also a firearms competitor is a good idea. After all, when applying lethal force against an adversary, it is critical that you hit what you are aiming at.

A successful competitor wins by hitting the target consistently, but this does not mean that he is the best person to become an instructor. The ability to perform any physical skill does not mean that a person can teach another to perform the same skill.

Although possessing a thorough understanding of the subject matter is critical, it takes more than that to be an effective instructor. A person must understand how others learn, be an effective communicator, have a desire to teach others, and be able to commit lesson plans and training records to paper in a comprehensible manner—and much more. Far too few competitors are effective communicators. Even fewer have the patience necessary to bring the substandard shooter up to par. Their attitude seems to say: "This isn't rocket science. Look how good I am. Why can't you do the same thing?"

It is also important to look at the type of competition in which the instructor participates. A popular competitive shooting sport is Police Pistol Competition (PPC), which is a good test of marksmanship and weapon handling but does not necessarily relate well to street situations. At the distances equal to the statistical norm for law enforcement shootings (7 to 21 feet) the competitors stand in one position, fire multiple rounds, reload, and fire again. The use of cover or the reduction of the competitor's silhouette is not allowed until target engagement at distances well beyond common engagement ranges. The adage of "as ye train so shall ye fight" has been proven time and again by catastrophic results to officers involved in lethal-force confrontations who choose to stand and return fire, rather than seeking cover and then returning fire.

Even when cover is introduced into the competition, it has little to do with practical applications of cover. The "barricade" is, instead of protective cover, simply a smooth post, which becomes a rest for the shooter's weapon. When the officer gets into the streets, little of the cover he will encounter is suitable for use in the same way. And when it is suitable (as in a streetlight post, for instance), it is for all practical purposes useless as cover. The more common cover available (e.g., automobiles, fire hydrants, trash containers) does not readily lend itself to support for the gun hand the way the so-called barricade does, and yet no PPC competition uses such realistic items.

As an example of what can happen when the officer is unused to using commonly available cover, an officer involved in a confrontation took cover behind a Dumpster. He leaned against it just the way he had leaned against the barricade in training, but the Dumpster was not fixed in place. It rolled away, causing him to lose his balance and fall, and then rolled back, trapping his left hand. Fortunately, the suspect ran away.

In PPC, little or no emphasis is given to keeping anything but the feet behind the barricade. Thus, when officers do take cover during a confrontation they all too often expose valuable parts of their bodies as targets of opportunity for the adversary.

International Practical Shooting Confederation (IPSC) competitors scoff at PPC as totally unrelated to actual law enforcement engagements. However, even though the original idea of IPSC shooting was to develop a more practical style of competition, the boasting is not completely justified. Granted, there are a variety of types of cover, moving targets, and other challenging problems (such as shooting from a stair or through a window), but there are still problems. Since winning is based on high scores and low times, sound tactics come in a poor second. In an IPSC match, such maneuvers as leaving cover with an empty weapon and reloading on the run are quite common. IPSC targets don't shoot back, so these lapses in sound tactics bear no punishment. The ultimate punishment, of course, comes when such tactics get the officer killed or wounded on the street.

That being said, being a competitive shooter should not disqualify a person from selection as a firearms instructor, but neither should it be a prerequisite.

Even when good potential instructors are picked, they are often shorted on their own training program. Many state law enforcement regulatory agencies consider the completion of a 44-hour firearms instructor's course to be adequate training to certify an instructor. Some states require additional courses in methods of instruction, line-coaching training, or other associated subjects. Still other states require a certain amount of on-the-job training at a state-certified training site. Even in the best of situations, that's where the training ends. There is one serious fault with such training. The courses almost overwhelmingly concentrate on teaching basic marksmanship and have little or no instruction in developing practical, tactical training courses.

Not all firearms instructor training ends with the minimal certification requirements. Many agencies send their firearms instructors to additional courses, and many instructors whose agencies do not provide further training attend courses on their own, at their own expense. Most of this training falls into the category of "advanced" firearms techniques and is produced by for-profit training centers. While many of these centers provide excellent training, in too many cases the officers who attend do not get what they need to add to their bag of instructor tools.

These training centers are commercial enterprises, and few exist just to provide law enforcement training. In fact, few would survive if they depended solely on the law enforcement trade. They survive by attracting people with money who want to learn fast and fancy gun work—become a "pistolero," as some advertise. To attract this clientele, these facilities must produce gimmicks that may look good on a video, but have little application in the real world.

The primary problem with many of the techniques taught lies in practice. Most require an extensive amount of practice to master, and, as we have already seen, there is seldom enough time and money for the average officer to do this. With the limited time and budget that constrain the average firearms instructor, all training must be aimed at what the average officer can use and master. It does little good for a firearms instructor to learn techniques that cannot be taught to the average officer without intensive and lengthy training sessions.

MONEY

Money has the greatest effect because it influences all the other factors. Training is expensive, and of all categories, firearms training is one of the most costly. The big-ticket items are weapons and ammunition, but since the officers have weapons anyway, the first can be discounted. Ammunition is another story. The ammunition budget for a major police department (of which I have personal knowledge), an agency with more than 1,000 sworn officers, is less than $20,000 per year. This equates to approximately 80 to 100 rounds per officer per year. Officers are expected to buy their own ammunition to cover the shortfall!

Add to this the multitude of minor costs that seem small in themselves but inexorably build up: targets, staples, hearing protectors, cleaning supplies, range maintenance, and, of course, the salaries of the instructors themselves. One can readily see why the training budget is seldom adequate.

Although tax-cutting measures abound in this land, citizens have a tendency to demand more and more services from the government. Residents who live in communities that cannot afford even to repair police and fire vehicles demand that little-used streets be immediately cleared of snow. These demands have spread to law enforcement agencies. Thirty years ago many such programs as Drug Abuse Resistance Education (DARE), formalized crime-prevention units, and runaway intervention units would have been scoffed at. Now they are considered essential. The problem is that the same citizens who insist on these added services are not willing to pay the extra taxes to support them.

All too often, when they are faced with budget shortages, law enforcement administrators are hesitant to cut "social" programs because they are so prominent. Many of these programs were started with federal subsidies and cannot be cut even though they now use department money and resources. Unfortunately, when such austerity programs occur the first items to be cut are the "invisible" ones, the ones the public doesn't ordinarily see. Training is usually one of the first cuts, and firearms practice is the very first to go. Often, the rationale is that the average officer will never have to use a firearm during a career in law enforcement, outside of annual requalification.

That such measures are a false economy is borne out by the profusion of lawsuits filed after any police shootings. Damage awards for single incidents are huge, far greater than the training budgets would be for years to come. How much training could be done if the most recent $25 million settlement incurred by the major city referred to in the previous chapter had been spent on teaching its officers proper tactics and increasing their shooting ability rather than filling the pockets of tort lawyers?

Off the record, city administrators will tell you their seemingly cynical but very practical reasons for paying such awards. These are extra budget items, and the money for such payments can be obtained from outside the normal budgets. Under such a system, the lack of money for training is a factor that will get worse rather than better.

EQUIPMENT

Even agencies that do not suffer from severe shortages of money or time find that their training facilities limit their ability to perform realistic firearms training. We have described the inadequacy of firearms ranges modeled after the military to perform tactical training, but that is not all.

In the two cases cited in the preceding chapter (*Popow v. The City of Margate* and *Zuchel v. City and County of Denver*), the courts included the lack of decisional training in the reasons for their rulings. Historically, providing effective decisional training has proved either difficult, expensive, or both. One of the early commercial offerings involving decisional training was a series of films from Motorola entitled *Shoot—Don't Shoot*. The films provided a variety of scenarios during which the participating officer had to decide whether or not to shoot at a suspect. Compared to current decisional shooting technology, these films were crude, but they were a start.

Today, there are CD video player devices that project situations and then critique the participant's decisions, actions, shooting ability, and skill. The technology has progressed to the point that the scenarios react to the actions of the student. One of the latest offerings will even "shoot back," firing plastic projectiles at the student if he doesn't use cover correctly and offers a body part as a target. A "pain penalty," if you will.

As good as this equipment is, there are some drawbacks. First is expense. Many systems are beyond the means of most agencies. Only larger departments, regional and state academies, and some successful private training facilities can afford the $80,000 to $100,000 or more that the equipment costs.

Second, some of the trainers who operate the equipment forget or ignore the fact that the purpose of the equipment is decisional training and turn it into an extension of the firearms training program by making the bulk of the scenarios involve shooting situations. The sales personnel who demonstrate the equipment encourage this concept by almost exclusively using shooting situations to show the features of the equipment because those situations draw the most immediate interest. Overemphasized exposure to shoot-only situations tends to make an officer expect every situation to involve lethal force, something that is certainly not encountered on the street.

Another deficiency of this type of equipment is that it cannot respond automatically to proper authoritative commands given by the officer. It is at the mercy of the operator, who has to make a decision on whether or not the verbal commands are effective, and who has the option (on the better systems) of changing the scenario to reflect that effectiveness. If the operator is not well trained, or does not respond correctly, the officer ends up not expecting compliance with his commands. It trains the officer, however subliminally, to expect his orders to be ignored and to respond to, and with, the use of lethal force.

Finally, this type of equipment has one major flaw. For all practical purposes, the officer is fixed in place, able to move only slightly in lateral directions to take advantage of cover. His attention is fixed on the screen directly in front of him. "Tunnel vision" is an almost inevitable result. Seldom do street confrontations occur in this way. The adversary may come from any direction, the officer may have to advance, retreat, or find cover somewhere else. Thus, although video systems offer a valuable training tool, they are only one portion of an effective training program.

To address these concerns, many police agencies have attempted to engage in force-on-force training. It is only when you are going up against an adversary who can plan his own actions, react to those you take, and stymie all your carefully laid concepts of operation that you can get serious training done. The military has recognized this for a long time. Military exercises have involved force-on-force training in the form of field exercises since well beyond my memory (and that is a very long time).

Traditionally, military exercises have been conducted using blank-firing weapons, each maneuver unit equipped with an umpire who makes the final decision on who won and who lost. The problem, also recognized early by the military, is that such exercises too often degenerate into "I shot you first"—"No, I shot you first!" arguments.

Various systems have been used to eliminate this problem, the most successful probably being the MILES system, in which each individual/vehicle/weapons system is kitted out with a series of sensors, which react to infrared or laser emitters affixed to the weapons of the enemy. No question of who shot whom—or when—under such a system.

The MILES system, however effective, is well beyond the financial means of most police agencies. It requires not only a heavy initial outlay but also continuing maintenance. How, then, to conduct realistic force-on-force training?

A solution is in paintballs. They are used perhaps less than they should be because their use is time intensive and can be quite expensive. The cheap paintball guns available at most discount stores have little in common with service weapons, and training officers often fear that this will negate the value of the training. And while paintball's offspring, using specially designed marker cartridges fired from modified service weapons, helps to offset this effect, such systems are quite expensive in the initial setup. This problem becomes particularly acute when an agency allows officers different types and brands of handguns, each type then needing modifications. The ammunition most commonly used in these systems is also relatively expensive, in many cases reaching a dollar per round.

Even when paintball or marker-cartridge training is conducted, the scenarios are often far too loose and result in chaos. There are two things that typically go wrong in the training session, and they are directly attributable to lack of management.

The first is the degeneration of the force-on-force training session into a paintball game. Participants fall into having fun rather than taking the session seriously and exhibit tactical behavior that is completely unrealistic. I have observed participants in such sessions charging across an open area, betting that they can "dodge the bullet" until they can get in a hit of their own. Early marker-round training had a "pain penalty" attached, the impact of the projectiles producing a nice welt, but with a few injuries and the threat of lawsuits, this penalty has gone away. Now the student is equipped with so much protective gear that the impact is barely felt, if at all.

To eliminate the "gaming" aspect, scenarios are often set up using either instructors or members of the tactical team as the adversaries. This can result in the second training problem, the "no-win" situation. Rather than key the responses to the skill level of the trainee, the adversaries turn it into an episode of "kill the cop." The skill level of adversaries being far higher than that of trainees, the latter inevitably lose. Frustration goes up, trainees become gun shy, and a negative training experience results.

As with video training, there is a final flaw to this training: it becomes the be-all, end-all of the training experience. Officers who receive their tactical training only with paintball or marker-round systems do not experience the final, essential "mind shift" that they get when they use their service weapons, firing live rounds, in a tactical situation. It is an almost intangible thing, but readily observable in trainees. Instructors at our training facility remark on the inevitable change when the trainee, often highly skilled in practice tactical maneuver, enters the live-fire shoot house for the first time. If the thought processes of the trainee could be put into words, it would be, "Damn! This is serious. I could get myself killed or kill someone else if I screw up."

A training problem that also rears its ugly head is the lack of specialized training for anyone except the tactical team. Ask narcotics officers where most of their work is done, and one likely answer is, in bars. Yet they are seldom if ever taught to draw and fire their weapons from a seated position. Are motorcycle or mounted officers taught to fire from the saddle of their bikes or horses? Do K-9 officers participate in firearms training with their animal partners present and ready to assume their own tactical role, and are other officers who may be on the scene taught what to do and what not to do when a trained dog is present? If officers are allowed (or required) to carry concealed weapons off duty, are they taught the proper methods of drawing from under a coat, from an ankle holster, fanny pack, or wherever else they choose to carry the weapon? Too often the answer to any or all of these, and many more, is no. Ask why, and you will get TIME.

Is it any wonder, then, that according to U.S. Department of Justice statistics, 84 to 85 percent of all police bullets *miss* their intended target?

A solution, already widely implemented, is to change police handguns from revolver to semiautomatic. With more rounds in the weapon—so the theory goes—we have a better chance of hitting the target. True. It is also true that the more rounds you fire, the same proportion will miss the target, meaning more expended rounds downrange, too often hitting someone with an extremely expensive lawyer. As an old team sergeant used to tell me, "You can't miss fast enough to win a gunfight."

I have spoken before about how tactical teams get a disproportionate share of what tactical training there is and the lion's share of specialized equipment. And while it is true that these teams are designed to go into harm's way, we tend to forget that it is the street officer who constantly faces high-risk situations—and generally without the benefit of an operations plan, surveillance, rehearsals, and all the other benefits of knowing what the situation is before having to go into it. By far, the street cop takes the most casualties in any police department. By far, the majority of incidents of "bad shoots" is attributable to the street cop. And, unfortunately, for most agencies this is the way it has been for at least the last 30 years, and this is the way it is likely to remain. The officers are given repetitive marksmanship training with few practical skills attached. The first time they fire their weapons in a tactical situation is when an actual tactical situation occurs.

Thus the law enforcement officers and the communities they are sworn to protect are put into an undefendable situation. Such cases as *Popow, Zuchel,* and *Allen* will continue to ruin the lives of officers and citizens.

CHAPTER 7

Earlier Tactical Training Programs

Modern-day tactical training facilities are the direct offspring of military facilities that sprang up in the late 1970s, all of which were a response to the spate of terrorism and hostage taking that were features of the international scene. The military quickly realized that the normal training undergone by soldiers was wholly inadequate to the task of surgically removing terrorists while keeping hostages alive. Incidents such as the debacle at the Munich Olympics in 1972, where German "sharpshooters" attempted to eliminate the Black September terrorists holding Israeli athletes hostage and got everyone killed in the process, drove home the lesson.

In the decades before this, at least in the U.S. military, training had concentrated upon defeating the Warsaw Pact hordes as they came through the Fulda Gap in Germany. For the infantryman, this meant forming pockets of resistance, usually centered on some antitank weapon, that would at least slow down the tide. What was not often said, but what was on everyone's mind, was that this situation would last only until someone threw the first nuke, whereupon all would escalate into a full-scale nuclear war and everyone would die.

Our ill-fated involvement in Southeast Asia quickly showed, at least at the level of those actually fighting it, the inadequacy of European war training. Tanks were for the most part useless. The enemy didn't have tanks, at least in the early years, so antitank weapons just became something heavy to carry around. The distances of engagement were extremely short, measured in tens of meters rather than hundreds. The M14 in 7.62mm NATO round, with which all units were equipped, was quickly seen to be "too much gun." Ideal for shooting long distances on the plains of Europe, and with supply channels adequate to the task of keeping the heavy ammunition close to the battlefield where it would be needed, the M14, like its predecessor the M1 Garand, was perfect for the task of conventional warfare. Its replacement, the M16 in 5.56mm, was far easier to control on full-automatic fire, and the ammunition was relatively light, enabling the infantryman who might not get resupplied for quite some time (if at all) to carry enough to perform his mission. And if it couldn't shoot out to 500 meters? Didn't seem too important in an environment where the target was far more likely to be at 5 meters rather than 500.

We also quickly realized that the tactics so appropriate to the European battlefield were inappropriate, to say the least, for jungle warfare. Unless you counted sitting on the perimeter of a firebase, there was no front line. Seldom did the enemy occupy some fixed position known to you where he could be attacked. The predominant tactic became patrolling, where you had to go out and find the enemy—

search—before you could fix him in place and eliminate him—*destroy.*

By their very nature, patrol engagements are very short, sharp, and fought at close range. There simply isn't time to acquire a good sight picture, apply lead if the target is moving, squeeze the trigger, and follow through. By the time the U.S. infantryman did all that, the engagement was often over with—in the worst-case situations with American dead and wounded and little else to show for it.

Recognition of this fact led to new training. When I returned from my second trip to Vietnam in 1967, I was assigned to the Weapons Department of the U.S. Army Infantry School at Fort Benning, Georgia. I became one of the earliest instructors in a program then known as "Quick Kill." This effort, first enunciated by an Olympic-class shotgunner, taught the shooter to ignore the sights (in fact, sights were taped over) and to shoot instinctively, just as if hunting quail. Students were issued BB guns, specially made by Daisy with no sights along the barrel, and were drilled in aerial target engagement. They quickly learned that you could not think a shot through, could not track or lead the target, and could not apply trigger squeeze or any of the other hundreds of refinements worked on by generations of marksmen—and still expect to hit the target. There simply wasn't time—as there wasn't time to engage a Vietcong soldier popping up from a spider hole, firing a burst from his AK-47 and disappearing.

After shooting thousands of BBs at aerial targets of ever-decreasing size and at scaled-down ground targets, the students made the transition to the service rifle, with which they applied the same techniques. Most became adept at close-in target engagement. In fact, the Quick Kill program was credited with vastly increasing the kill ratio for U.S. infantrymen in Vietnam.

The program had its detractors, of course. The

cadre at the Fort Benning Marksmanship Training Unit (MTU), whom we referred to as the "Camp Perry Bunch," were horrified. We were teaching the soldiers bad habits, they said. What would happen when the infantryman had to engage targets past a hundred meters? Would Quick Kill work then?

The answer was, of course not. As for teaching bad habits, who knew? What we did know is that the average infantryman would never see the enemy any farther away than 25 meters, and seldom even that much. A truce of sorts was negotiated, with the Quick Kill adherents continuing to train the line grunts and the MTU training the snipers.

Of course, with our disinvolvement in Southeast Asia and concurrent reemphasis on the battlefields of Europe, the traditionalists triumphed. Quick Kill slowly died and is being taught nowhere in the military at present.

I bring this issue up only to illustrate a couple of points. The military, confronted with a situation for which traditional training is inadequate, can and will adapt. But perhaps the most important point is something that we in the Quick Kill committee recognized early on. The methods taught in the program were probably less important than the fact that the students got to shoot thousands of rounds instead of the hundreds they would shoot in traditional marksmanship training. Practice really does make perfect, or a reasonable facsimile thereof.

The adaptation process was certainly shown in the military's reaction to its new mission in the 1970s, anti- and counterterrorism. After Munich and a host of other incidents, the U.S. Army realized it was going to have to have certain well-trained personnel who could, at a moment's notice, move to a hostage-taking situation and resolve it, with—it was hoped—a minimal loss of life. Almost as an ad hoc response, a program was begun at Fort Bragg at the U.S. Army Special Warfare Center that became known as Blue

Light. Special Forces A teams became, almost overnight, the antiterrorism response units for the military.

This meant training, and among the Special Forces, practically all of whom at that time were combat veterans, it meant live-fire training. The problem was, how could you train to take down an aircraft or search a building using full live fire? There were plenty of old unoccupied buildings at Fort Bragg, particularly since the post-Vietnam drawdown. Should we just go in and shoot them up? Safety concerns, of course, quickly precluded such courses of action. Some specialized training facility was needed.

MICHELIN CITIES

The requirements for a live-fire shoot house were simple. It had to be sturdy, it had to keep all the rounds inside the facility, it had to eliminate the problems of ricochet and splashback, and it had to be cheap. The military in the 1970s wasn't getting a lot of money, and the Special Forces had traditionally been the poor stepchild even in eras of plenty.

In a pattern that would be repeated many times in the future, we turned to our allies, who had been dealing with the problems posed by terrorism far longer than we had. In particular, we looked to our sister unit in the United Kingdom, the Special Air Service (SAS), with whom we'd exchanged liaison officers and noncommissioned officers for many years.

From the SAS came the idea of using rubber tires filled with dirt or sand, aptly named "Michelin cities." The tires were stacked in configurations meant to simulate the rooms and hallways teams would encounter during an actual rescue attempt. On the face of it, this seemed an ideal situation. Bullets fired into the tires would be absorbed and finally stopped by the resistance of rubber and sand. Ricochet danger was slight, as was that of splashback. Even high-powered rounds in 7.62 NATO could not fully penetrate the barrier created by the sand-filled tires. And, of course, it was cheap. Tires could be had for nothing, the post engineers could be badgered into bringing loads of sand to the site and dumping it, and in that era at least, Special Forces soldiers were accustomed to doing all the heavy lifting themselves.

It was a workable solution (and one still used in many places). Thousands of hours of training, and perhaps hundreds of thousands of rounds, were expended in the large facility built at the Special Operations Training Center at Mott Lake at Fort Bragg. Until Special Forces Operational Detachment DELTA was formed and trained, the Blue Light teams drawn from the Special Forces groups at Fort Bragg were the primary response force for terrorist incidents and kept their skills at a keen edge. Even after Blue Light stood down, the facility was still used in special operations training: the Special Warfare Center recognized that it would be a good idea to train every Special Forces soldier in these essential skills.

That heavy use, however, quickly revealed the problems inherent in such facilities. Unless the targets were moved frequently, the multiple bullet strikes in a fairly small area quickly broke down the tire walls, allowing sand and dirt to leak out. When enough sand was gone, the wall was no longer ballistically sound, thus producing the danger of shoot-through. In fact, several soldiers were wounded when a burst of submachine gun fire penetrated one such wall.

Even worse, the soldiers soon started to experience a problem with ricochets and splashback in a facility that had been designed to eliminate such dangers. Investigation showed that even in the tires that had not been shot often enough to degenerate, there was still a problem with the rounds lodged in them. The bullets were ricocheting off buried bullets that had been fired before. And it could not be determined just which tire had enough bullets in it to cause this problem, which meant tearing down and completely reconstructing the facility quite often.

Another problem was training "burn-in." Inasmuch as the walls of the facility could be moved only with great effort, the rooms and hallways were mainly kept at the same size and configuration. And this forced instructors to recognize a phenomenon that at first amazed them. If you want someone to completely imprint every possible detail of a room into his mind —the "burn-in"—send him though the room using full live fire. Changing the inside features of the room and moving targets around are of only limited use. The trainee quickly becomes so accustomed to every detail in the facility that he could negotiate it in his sleep. And sending him through again and again only produces overconfidence and boredom.

As for realism, even covering the tires with plywood and painting it to more closely resemble an actual room was also of limited use. The place still looked like a tire house. The roof was open to the elements, and so the lighting could not be manipulated. After all, seldom does a hostage rescue take place in bright sunlight, and yet that was the way the soldiers were being trained.

But what ultimately drove the final nail into the coffin was the same problem I have spoken about for conventional ranges. All those lead bullets fired into the tires? Hazmat. And a huge remediation process. Now not only do we have to get rid of the bullets and the dirt, but the tires as well. And the normal process for remediation—incineration—does not work well for used tires because it clogs up the incinerator. Want to guess how much that one cost?

RAILROAD TIE HOUSES

Shortly after the advent of Michelin cities, another technique was tried. Used railroad ties were readily available, usually for free: they were far easier to stack than tires; and you didn't have to fill them with sand. The wood itself would stop the bullet, and you could eliminate the problems with ricochet and splashback as well. A facility constructed of railroad ties was also somewhat more analogous to actual construction and, if covered with a roof, was amenable to appropriate lighting systems.

In practice, however, railroad ties were less than satisfactory. The ties were of course used. Barring obvious rot, that meant that you couldn't be sure what you were getting. The tie might be whole, or it might be a shell covering a hollow interior that was obviously not ballistically sound. You might be able to spot a hollow tie by weight, but you could not do so if only a small portion of the tie was rotten. Short of X-raying each tie, you could not be sure that a round fired into a certain location would not bore through the tie.

Even though you didn't have to worry about sand or dirt leaking out, you did have to worry much more about shoot-through. Placement of targets had to be changed practically after every iteration, and even then sooner or later you were going to have to replace some ties.

The imbedded rounds in the ties, and the consequent ricocheting of new rounds off them, was just as big a problem as with Michelin cities. Further, rounds fired at acute angles, especially large, slow rounds (as are most .45 ACP), would ricochet off the tie itself.

Finally, the hazmat problem once again reared its ugly head. Here are all these ties with thousands of rounds in them—what are you going to do with them?

PURPOSE-BUILT FACILITIES

We tried various other methods. At one point we built a city of target cloth and 2 x 4s in the middle of the desert to train a special battalion of the Saudi Arabian National Guard. As long as only one unit at a time went through, command

and control was tight; and if we could keep the camel herders out of range, it worked well. It is easy to see how poorly this solution would work practically anyplace else, however.

The British were, once again, ahead in this field, partly because of sheer necessity. Northern Ireland had devolved into the killing zone from which it has only recently (it is hoped) emerged. British squaddies were confronted with urban warfare, with all its inevitable complexity: where to kill or wound an innocent was a surefire way of strengthening the other side—where targets popped up from windows, sewers, hallways, and streets; fired one or two rounds; and disappeared.

Whole city streets, then, had to be constructed in which to train. Obviously, the tire houses, railroad tie facilities, and all that had gone before were inadequate to the task.

The photos on the following page show what they came up with. The street could be one drawn from any town or city in Northern Ireland. Targets pop up in windows or doorways, or down the street, and disappear. The targets could be moms pushing prams, two old men turning a corner, a housewife opening the window to get some air, and any number of other citizens going about their business. They could also be hooded gunmen behind the sights of sniper rifles, grenade-throwing hooligans, or a bunch of kids chucking rocks at the hated Brit soldiers.

Every unit assigned to Northern Ireland has to go through this training, which is credited with saving perhaps dozens of lives, both British and Irish.

The material that made this facility possible is called UTILEX Ballistically Sympathetic Concrete. This material, lightweight and yet strong, was originally invented as a solution to the problem of constructing cost-effective and energy-efficient walls for low-cost housing. Quite by accident, it was discovered that a bullet fired into the material would be absorbed and held instead of ricocheting and spalling, as is normal with standard concrete. A building constructed of

These British training facilities simulate conditions in Northern Ireland and provide live-fire training for British soldiers.

this material looks and acts like any other building, except when hit by a bullet or other high-speed projectile.

The only problems associated with UTILEX facilities are cost, and upkeep. The Combat Village built in Hong Kong in the mid-1980s (shown on page 80), which covers 7 acres, cost £7 million (approximately U.S.10 million at current exchange rates), exclusive of any earthwork and site preparation. Although this is quite cheap for what the security forces of Hong Kong got, it is well beyond the means of most police departments.

Upkeep consists of replacing the concrete panels when they have absorbed a certain amount of ammunition. The upkeep can be kept reasonable by frequent target placement changes, and the replacement process itself is relatively simple but has to be figured into the maintenance of the facility. Otherwise, just as with tire houses, railroad tie houses, and so forth, there can be shoot-through or ricochet/splashback from the embedded rounds.

In the United States we took a different approach. Our emphasis was less on urban warfare and more on small-scale antiterrorist operations, and thus it was decided that we didn't need facilities of the scale of the Hong Kong and United Kingdom/Northern Ireland training sites.

Our facilities were truly "shoot houses" (it was decided early on that the term "kill house" was far too bloodthirsty and politically incorrect). The training house at Fort Bragg, built for DELTA, is typical of the breed. First, a steel infrastructure was put up. Then armored steel panels (generally a half-inch in thickness and of Brinell 500 in hardness) were bolted or welded onto the steel frame. These panels had to cover the walls, floor, and ceiling. Then very thick rubber anti-ricochet panels were glued onto the panels. Finally came installation of ventilation and filtration systems to rid the facility of toxic fumes. On the whole, such a facility consisting of only a few rooms, with perhaps two floors, cost as much as the entire

THE CASWELL DEFENCE COMBAT VILLAGE IN HONG KONG WAS MADE OF UTILEX BALLISTICALLY SYMPATHETIC CONCRETE.

Hong Kong facility. Such a place was, by necessity, impossible to change once it was built. And, sadly, as demonstrated by a couple of incidents, there was an inherent flaw, at least when standard ammunition was used. The rounds would, of course, be embedded in the rubber anti-ricochet panels. The spent bullets being very hot from the firing process itself got even hotter if they hit the underlying steel plate with any force—hot enough to set the rubber on fire. This was seldom noticeable at first; the rubber just smoldered for hours before achieving enough heat to set the adjoining panels on fire. But once the fire was under way, it was practically impossible to put out. The heat, of course, would remove some of the temper from the armor plate, and the process had to be started all over again.

And to a degree even greater than those employing UTILEX, the cost of such facilities was well beyond the reach of even the richest of police agencies. Since the facility at Fort Bragg went up, there have been other similar shoot houses constructed, some more grandiose, some much smaller. Reports from the Federal Law Enforcement Training Center state that its shoot house is so small that "it's like having a shoot-out in your closet." Costs are also obviously a large factor.

RANGE ADAPTATIONS

Those police agencies that attempt to build tactical training facilities usually do so with an adaptation of an existing range. The reasons for this are obvious. The land is already there, and buying or obtaining land through eminent domain takes time and generally more money than the agency has to spend. Neighbors are already accustomed to the sound of gunfire; though they may complain about it, this probably won't be any more than usual. The facility infrastructure (target houses, administration buildings, weapons lock-up, ammo storage) are already in place, as are parking, access to and from the site, and fencing.

However, unless there is a lot of unused space on the site, a serious problem rears its ugly head. Most police ranges are already in use (or overuse) for a large part of the year. Not only is the police department that owns the range going to be using it, but so are a host of other agencies. Time is already at a premium, so setting aside one portion of the range for strictly tactical training is problematic. Any tactical training facility to be used on existing firing lines has to be easily portable and removable, without degrading the capability of the standard range. Adaptations include using plywood or fiberboard panels that can be hooked together to simulate rooms and hallways or constructing wood frames to be hung with plastic or cloth. These adaptations can only be used where berms rise to at least 8 feet all around the facility, since the flimsy adaptations are obviously not going to stop any rounds. And because few ranges have berms completely surrounding them, the tactical maneuvers will have to be carefully controlled, so as not to shoot into unprotected areas.

Even those ranges that do have extra space, where a "shoot house" made from commonly available materials can be set up and left in place, suffer from command and control problems. They are certainly sufficient for two- and four-man teams, where everyone is going to be together during the entire training scenario, but do not readily lend themselves to greater numbers of trainees. There is far too much danger that one team will shoot through a wall into a room where another team is training. Such exercises can be controlled by rigid rules and constant supervision, but this is somewhat unrealistic. There will be no strict control by an instructor of a tactical team going through hallways, stairwells, and rooms of a business or school in an active shooter situation.

OTHER ATTEMPTED SOLUTIONS

I know of a couple of departments that actively seek condemned property, preferably out in the countryside, where they can go in and practice entry and clearing. Aside from the problems of shoot-through, have you ever considered the number of substances in the walls and ceilings of an ordinary building that, when hit, are likely to deflect rounds into places you don't want?

One agency tried to use an old rock quarry. After firing just a couple of rounds and hearing them zing past on their way in ricochets, that idea was abandoned. Still other organizations have attempted to modify existing buildings, particularly those built of concrete, by hanging plywood panels on the walls to deflect ricochets. Paradoxically, this worked very well when the entry weapons used were in 5.56mm. This round, traveling at 3,250 feet per second (fps), actually started to break up on the plywood panels and seldom penetrated both panels and the concrete behind. Pistol bullets, on the other hand, worked like jackhammers, quickly compromising the integrity of the walls.

So is it any wonder that little realistic, live-fire tactical training is actually done, even for tactical teams? And as I said before, even less is being done for the street cop. Is there a remedy for this situation?

I think so.

CHAPTER 8

Nontoxic Frangible Ammunition as a Solution

At the beginning of the book, it was asked if NTF ammunition might be a round for all reasons. The answer is easy: of course not. However, it does present some interesting solutions to problems faced by firearms trainers.

First, let's consider the lead problem. NTF eliminates it. Period. End of story. There is (or should be, and we'll take that matter up in a moment) no lead in NTF ammunition, either in the primer or in the projectile.

There should also be no other dangerous heavy metal. After all, it wouldn't do much good to replace lead with cadmium, which is far more toxic. When ordering NTF, ask for a Materials Safety Data Sheet (MSDS), an official form that lists *all* the materials in the product that might pose a health hazard, no matter at what level. Insist on the MSDS for the primer, projectile, and powder. Inspect the MSDS for the following materials, making sure that if they are present they are at or below the levels listed with each. (NOTE: The following list of toxic materials is taken from an FBI request for proposal for nontoxic ammunition.)

The Ammunition Shall Not Contain:

1. Any trace of lead or mercury
2. In excess of ambient levels, or 0.05% concentration by weight of all below materials:

   Barium
   Beryllium
   Antimony
   Cadmium
   Arsenic
   Chromium
   Selenium
   Tin
   Thallium

This list was, by the way, challenged by one manufacturer that stated it was not aware that tin was a toxin and asked that it be removed from the list. To which the FBI replied:

> According to the American Conference of Government Industrial Hygienists, elemental tin has low toxicity, but most of its compounds are toxic. Exposure to tin, especially tin dust, affects lung tissue. This occurs through inhalation and is a result of long-term exposure. The concern involved is sufficient enough to include tin as a toxic material.

But even before securing the nontoxic ammunition, agencies that wish to entirely eliminate lead in indoor ranges want to first remediate the range to get rid of the lead that has already been accumulating. This should be done professionally, and, if it is an indoor range, the costs are not *too* onerous. All the lead and other expended materials have to be taken out of the bullet traps, of course. All surfaces that might have collected lead dust must be washed down with ordinary soap and water. The filters in the ventilation/filtration system must be removed, and, if disposable, disposed of. Otherwise they must be thoroughly cleaned.

Even with all this, the staffs of ranges that have undergone this process are often surprised to find upon conducting air-sampling tests that lead is still being produced in the firing process. This is not necessarily a sign that someone has phonied up an MSDS. The effect is particularly pronounced if a lot of revolvers are being fired, particularly if the shooter has been firing solid-lead slugs (wadcutters and the like). Even with a thorough cleaning, there is still a coating of lead in the barrel and down in the rifling. That lead is still produced while NTF is fired means that the ammunition is actually cleaning any remaining lead from the weapon. This effect will go away after a time. (If a shooter continues to use lead rounds for off-duty shooting, the effect will not go away.) However, even if all the shooters on the line have been firing wadcutters all the day before and are now using NTF on a range, the effect of the stripped-out lead will never come close to the maximum permissible exposure level.

Are there manufacturers out there that claim their product is lead free when it really isn't? I've not heard of any, but that doesn't preclude the possibility. The smart course is to continue air sampling after the new round is used, and if any more than the trace elements referred to above are present, it might be worthwhile to send the rounds out for analysis. The mass spectrometer available in most crime labs will do an admirable job of identifying any compounds in the primer, projectile, or powder.

Outdoor ranges require less cleanup. The rounds hitting the lead already in the berm will not stir up enough dust to seriously damage air quality, and the lead dust that remains on the ground and in the grass will probably not be disturbed so much as to require remediation. If you have not already received a sanction from the EPA or its state-level equivalent, you can probably make the changeover immediately and eliminate your lead pollution problem simply by firing only NTF.

If the product is to be used in a shoot house, it is probably worthwhile to at least scrub down all exposed surfaces with soap and water, particularly if the facility is enclosed. This will largely eliminate the contamination problem but won't eliminate the danger of shoot-through in a tire house/railroad tie house. I'll take that subject up when I discuss the frangibility portion of NTF.

The problem, of course, comes when trying to change over completely to NTF. Few facilities will have the luxury of doing so, largely because duty ammunition is likely to continue as the leaded variety, at least in the near term. For purposes of limiting liability, most agencies are going to require that the officer fire in qualification what he is going to carry on the street.

So the range may not become, or remain, totally lead free, but the use of NTF for all training other than qualification is still going to drastically lower the levels of toxic materials on the range. This also lowers the costs of doing business: the filters on the ventilation system will be replaced far less often, the worries about blood-lead levels for instructors and range personnel will diminish, and the possibility of carrying lead pollution home will be drastically reduced.

For those ranges that do switch over completely to NTF, there is an additional saving. The material that collects in the bullet trap is also completely nontoxic. It can be disposed of as you would any other garbage, instead of having to use hazmat procedures and regulations. In fact some ranges, particularly those that use a copper-based frangible, now sweep up the material and sell it to scrap-metal dealers, who melt out the polymer typically used as a binder and recover the copper content.

On outdoor ranges, the material that collects in the berm, once the lead has been removed, can remain there indefinitely without worrying about contaminating the groundwater. What little copper or bronze that leaches out will taint the water even less than the normal electrolytic process of drinking water going through copper pipe.

Finally, even in outdoor ranges where a mixed lot of ammunition is fired, both NTF and leaded, the range maintenance problem is somewhat eased because a far smaller amount of lead will lodge in the berm. This cuts down not only on pollution but also on that problem I spoke of earlier: the probability of splashback and ricochet from bullets hitting a semisolid mass of lead and copper in the berm.

YOU'VE GOT NONTOXIC: DO YOU STILL NEED FRANGIBLE?

Some agencies will stop right there—the elimination of toxic materials being their first (and sometimes only) priority. This tendency is particularly pronounced if all an agency is going to do is punch holes in paper targets. Who needs frangible for that?

We'll consider the versatility of frangible ammunition later on, but right now I will list some of the reasons that frangible can help an agency cut costs and provide for better training, even by punching holes in paper targets.

Range Damage

Put quite simply, frangible ammunition damages ranges and equipment far less than conventional ammunition, or even nontoxic nonfrangible ammunition. This is not, as some might think, because of the relatively lighter weight of the projectile. In fact, to obtain the chamber/port pressures necessary to operate most modern autoloaders, most manufacturers of frangible ammunition will push the projectile at a much higher rate of speed than is the case with conventional ammunition. With the equation cited earlier, $E = mc^2$, there can be more foot-pounds of energy produced with frangible than with conventional ammunition. As an example, consider the following: a frangible 9mm projectile weighing 90 grains, traveling at 1,450 fps (relatively normal for this kind of ammunition) produces 443 foot-pounds per second at the muzzle, while a conventional 115-grain projectile traveling 1,120 fps produces 345 foot-pounds per second. The relatively higher energy is maintained all the way out to 30 yards, past the normal engagement ranges of most pistol training.

Yet frangible ammunition causes less damage to the backstop (and anything else it might hit on the way), than does conventional ammunition. Why?

The answer comes with the study of what the differing bullets do upon impact. Conventional ammunition remains largely whole, and thus the energy it contains is transmitted almost in its entirety upon the point of impact. This is why most range backstops are slanted, basically providing for a controlled ricochet, the striking round imparting only a portion of its energy on the initial point of impact and expending the remainder in bouncing around in a swirl chamber or striking other rounds already in the bullet trap. A 90-degree backstop wouldn't last very long under the pounding of conventional ammunition even if you could build one that somehow eliminates splashback and ricochet.

Frangible ammunition, on the other hand, is designed to come apart upon impact. Initial energy is expended on the point of impact, but the remainder is dissipated in the tiny particles that fly away and quickly lose their energy due to air resistance. A 90-degree backstop is not only feasible, it is recommended. As long as the metal is sufficiently thick to keep it from "bellying" with the impact of the rounds (usually 1/4 inch of armor plate or 3/8 inch of mild-steel plate), you can shoot pistol rounds at it almost indefinitely. I have one sheet of 1/4-inch Brinnell 500 plate that I have been shooting at for the past 10 years, and the only signs of impact are the copper smears polishing it to a high sheen.

Replacement of bullet traps becomes, if not a thing of the past, at least a chore that can be put off much longer. Other range structures (e.g., baffles, swirl chambers, target stands, walls) also suffer far less damage when frangible is used.

These features, in themselves, might not be enough to convince an agency to switch to frangible ammunition. After all, ranges are designed to take the beating given by conventional ammunition. A properly designed bullet trap will last for many years, far longer than most administrators are going to be on the job. Range maintenance is a continuing effort and probably already budgeted. Some other factor is needed.

VERSATILITY

Quite simply, many ranges use frangible because of the many things they can do with it.

Shooting steel reactive targets, for instance, was nigh impossible on a standard indoor range. Bullets hitting the targets defeated the purpose of the carefully designed bullet traps, in that the round never had a chance to glance off the surface of the trap, skid into the swirl chamber, and lose its velocity there. Instead, the round hit the steel and generally broke into several large pieces—most of which bounced around doing a tremendous amount of damage to baffles, walls, lights, and, all too often, shooters. Even when the bullets had lost all their energy, they were still a problem. They were no longer contained in the bullet trap but instead scattered all over the range, where they had to be cleaned up. Too often, the range maintenance personnel did not follow proper personal protection procedures while ridding the range of this detritus, thus increasing their chances of getting a lead overdose.

I've already spoken of the danger of shooting at a steel target at the normal engagement ranges of 7 to 21 feet. But even if the danger of splashback and ricochet could be avoided, there is still the problem of where to place the shooter and the target. Put the targets downrange near the bullet traps and move the shooters toward them, and the shooter is even more exposed to the lead wafting up from the bullet traps because most range designs do not allow for proper ventilation at those particular spots. Move the targets uprange and the damage suffered by the range structures is vastly increased.

Properly designed and manufactured frangible ammo allows the engagement of steel targets at near point-blank range, with little danger of ricochet or splashback. That's not to say that it absolutely cannot happen. Shoot a badly wallowed-out and scarred steel target with a frangible round, and it is liable to throw tiny fragments of the frangible back at the shooter. All steel targets should be flat and smooth to eliminate this potential problem.

The ancillary damage, particularly to range structures, is also drastically reduced, though particularly such fragile items as lights can still be struck. You must realize that even with frangible ammunition the particles do not magically disappear. To show exactly what they do, place a cardboard box, bottom toward you, over a steel target and then shoot through it onto the plate. When you remove the box, you should be able to see a perfect pattern of sprayed particles corresponding to the face of the target, about a half-inch in thickness, chewing the interior surface of the box. The particles that go downward will hit the floor, probably losing what little bit of energy they still possess in the process. Those going to the sides travel in a flat plane until they lose energy or hit something, such as the walls of the range, where they'll do no damage. Those going up, however, can still maintain enough energy to shatter a fluorescent light if it is directly above the target. To eliminate this problem, simply move the target forward or back a few inches.

The only other proviso is that other shooters should not move past the line of targets while someone else is shooting at them. The side splatter, up close, still possesses sufficient energy to cause injury. The safety officer's job does not go away while using frangible, though it is greatly simplified.

With these precautions, the targets can be placed anywhere on the range that the tactical scenario demands. Engagement of steel targets at interview-stance ranges? No problem. Multiple attacker scenarios? Also possible. What about 180-degree shooting (or even more)? Possible with a few simple and inexpensive range modifications, about which I'll talk later.

The versatility of outdoor ranges is also vastly increased. Although these ranges have always lent themselves to tactical training far more than indoor ranges, there were still the problems mentioned earlier. Yes, you can move the targets, but you have to accept the possibility of ricochets over the berm. You could move the shooter forward, but only up to about 15 meters from the steel target, an unrealistic engagement range. Further, multiple-target scenarios suffered from the angle problem, particularly at close range.

Frangible ammo virtually eliminates these problems, though caution must still be used in target placement and engagement. Again, make sure that no one fires on a steel target while anyone else is exactly parallel to the target. The side splatter, though not lethal, still maintains enough energy at close range to cause flesh wounds and injury to unprotected eyes.

Inevitably there's this question: Does frangible ammunition have and retain sufficient energy to drop steel targets designed for standard ammunition? The answer is an unqualified yes, *at combat ranges.* Out past 50 meters for pistols and 350 meters for rifles, the round loses energy rather quickly. But if you are shooting targets at these ranges, frangible offers no particular advantage, and thus it has to be assumed that shooters would fire nontoxic nonfrangible ammo.

FORCE ON FORCE

Occasionally this question is posed: Can frangible ammunition be used for force-on-force training? The answer is an emphatic NO. Confusion has arisen perhaps because one manufacturer of frangible also manufactures marking-type ammunition for force-on-force training, and so the two types have become confused.

Frangible ammunition is capable of penetrating up to 26 inches of 10-percent ballistic gelatin. The rifle rounds tend to penetrate less, making them even more lethal because the entire energy of the round is dumped into the medium. The first time I fired a .223-caliber frangible round at a block of gelatin 4 inches square by 16 inches deep, the gelatin block jumped 8 inches off the table. Closer inspection showed that the expansion of the block had blown a hole in the 3/4-inch plywood table. It is my opinion that the strike of such a round anywhere in the human torso would result in a wound that would not be survivable.

Therefore, I emphasize that the training officer must stress to the highest degree that frangible ammunition should be used with the same care shown with standard ammunition, that it should never be fired at a living target unless the intention is to direct lethal force to that target, and that in some instances it may be even more lethal than standard ammunition.

That being the case, let's explore the uses of frangible ammunition in the training program.

CHAPTER 9

Nontoxic Frangible Ammunition in Training

BASIC MARKSMANSHIP TRAINING

Other than its nontoxicity, nontoxic frangible ammunition offers no particular inherent advantage over standard ammunition for basic marksmanship training. It will punch holes in paper targets just as standard ammunition does. It is, given relatively recent advances in the field, just as accurate as the ammunition commonly used in basic marksmanship training. It is often, however, somewhat more expensive than standard ammunition (particularly if an agency has the habit of choosing the least expensive ammunition possible).

It becomes advantageous in those locales that have closed their ranges because of environmental problems. When this happens the trainees are thrown to other ranges not yet closed, with the attendant scheduling restrictions such actions entail. Transportation often becomes a problem. Certainly, time is wasted traveling to a range that is at a distance, time that could more profitably be spent in training.

This problem has become particularly acute in those agencies whose ranges were built long before there was any worry about the toxic effects of lead. Older police ranges, military installations, and particularly National Guard armories and Reserve Officer Training (ROTC) ranges have been hit hard. It has gotten so bad that in some states the National Guard must travel, not to another range within the state, but the next state over to accomplish their annual requalification.

A conference for National Guard safety officers I attended gave some hint of the magnitude of the problem. Not only had they closed down all the ranges in their particular jurisdictions, but they had no intention of reopening them. The cost to kep them open would simply have been too high; the ranges had been built so long ago that to reengineer them to take into account lead toxicity would have cost more than the construction of a new range. They admitted that this created huge problems with readiness. And since, with the drawdown of the regular army, the National Guard has become an even more integral part of national defense, this could be expected to create problems for the nation as a whole. Yet nothing is being done about it.

They also admitted that many National Guard commanders wouldn't have agreed to reopen the ranges if they could have, since they now use the space for storage and offices. One would have to draw one's own conclusion as to the state of mind that would allow a commander to think this way, but that is well beyond the scope of this book.

The advantage of NTF in basic marksmanship training lies in the role this ammunition can play in reopening ranges—and for a fraction of the cost of reengineering an old range to handle standard conventional ammunition. Ambient lead, which will have long since settled out of the air onto exposed surfaces, must be removed. There are many professional hazmat remediation companies that can perform the task for prices that are far below those of other hazmat remediation (e.g., asbestos). If the bullet traps have been removed, they can be replaced with steel sheets of at least 3/8-inch thickness, treated to a hardness of at least Brinnell 500. Otherwise, the old bullet traps can be used after inspection to ensure that there has been no shoot-through or excessive weakening. Ventilation systems need be no more than those ordinarily used to suck out powder smoke.

With these tasks performed, a range can be reopened and the basic marksmanship trainees taught right at their own facility, saving time, increasing proficiency, and cutting overall costs.

What if the old range has been irrevocably lost—torn down, perhaps, or put to other uses during remodeling? Then new ranges can be built, at a fraction of the cost of a standard range. Any open space with enough room to accommodate the ranges the shooter will need will work. Filled concrete blocks for the sidewalls, a normal HVAC system, steel plates as described above for the backstop, and steel baffles to deflect shots fired toward the ceiling—and you have a range. Further, some range facilities companies have designed firing ranges that can be erected on-site and then taken down and stored when not needed. (A list of sources is included in the Appendix.)

BASIC TACTICAL TRAINING

While NTF can solve many of the problems associated with basic marksmanship training, it is in tactical training that it truly comes into its own.

Chapters 5 and 6 listed in great detail the problems now encountered in tactical training, which simply is not done to the extent it needs to be; not for police officers, the military, security personnel, or even the citizen concerned with securing his own safety. TIME is the problem.

Let us take a look at the factors time, instructors, money, and equipment, and see how the use of NTF ammunition can alleviate the problems.

Time

As stated in Chapter 6, there is seldom enough time to do all the things desired. This is true for police administrators, military commanders, and citizens. So many other tasks, hobbies, and requirements take up what little time is available. So if even more time is taken to accomplish firearms training, such training is necessarily going to suffer.

Therefore, the ability to perform tactical training on existing ranges should be of immense benefit. Entry-level training (e.g., from police academies or the military) can be considerably enriched by taking little more time than that already spent on the range. Instead of seemingly endless, mind-numbing, repetitive drills, the instructors can throw in street-level tactical maneuvers that are both far more interesting and pertinent than just punching holes in paper targets.

In-service training, too, can be enriched. Many ranges that have adopted NTF now use it in concurrent training. That is, while mandated requalifications are run on one part of the range, officers not actively involved on the line are over to one side, learning the proper techniques for the close-in engagement of targets, firing from behind cover, and basic entry and room clearing.

Instructors

NTF ammo cannot by itself improve the quality of instructors. What it can do, however, when used as a tool for instructor training itself, is to make instructors aware of the innovative uses for NTF. Unless they are completely hide-bound, instructors do not automatically reject innovation.

Far from it. Most are dedicated to their job, which is to prepare trainees for the situations they are likely to encounter on duty. Give the instructor the tools to improve the training program, and generally they will be embraced. I have found, while demonstrating ammunition for groups of instructors, that the most useful and innovative suggestions for its use come from the instructors themselves. It is not unusual for a group of instructors to go away and come up with an entirely new training program, one that directly addresses the problems faced day to day in preparing their students for the tasks they will be undertaking.

Money

We are talking about an ammunition that, as a rule, is more expensive than the conventional ammunition used at most training facilities. How, then, can we say that using NTF will save money?

First and foremost, NTF saves immensely in downstream costs. The U.S. Department of Defense has calculated that downstream costs resulting from the use of leaded ammunition add at least 30 percent to the cost of the ammunition because of cleanups, potential health problems for trainees and instructors, and future liabilities. Remember that according to federal rules, the government can assign liability to every owner/operator of a site considered hazardous, all the way back to the time the site was first contaminated, even though subsequent operators/ owners contributed nothing more to the hazardous conditions.

A 30-percent add-on seems high but is probably conservative. Range cleanup, with the ever-increasing number of restrictions and regulations, is a very costly business and likely to become more so.

New ranges are simply not being built, primarily because of this problem. Offering the agency the ability to construct a simple, fairly inexpensive tactical range or to modify a portion of the training area to accommodate tactical training can be an affordable solution to training problems simply by the use of NTF.

Still, we constantly hear the refrain that "it costs too much money! Why, I can get (name the round) for X dollars a thousand, and this stuff costs Y! My budget can't take it."

All true. But aside from downstream costs, look at the actuality of the situation. How many rounds are going to be fired per trainee in basic tactical training? Far fewer than those consumed in basic marksmanship training. In the basic entry and room-clearing class conducted at a local training facility, each student fires fewer than 50 rounds of NTF during the entire course. Obviously, such costs add up—particularly if there is no basic tactical training at all, as in most agencies. But in the context of what can happen if the trainee gets no realistic training and is then involved in a lethal-force encounter, such as described in earlier chapters, it can be seen that the costs are affordable.

Equipment

In using NTF, the equipment needed for tactical training is easier to make, cheaper, and often more versatile than that needed for use with conventional ammunition. In a later chapter I'll discuss in detail ranges, targets, and devices that can be used to enhance the tactical training experience. Suffice it to say, it's easier than you might think.

ADVANCED TACTICAL TRAINING

Today's advanced tactical training programs are mostly for SWAT teams, military special operations units, and highly skilled civilian security organizations such as those employed by the Department of Energy. We do not train the street officer or patrolman, the squad-level infantryman, or the armed security officers so ubiquitous in today's society.

We cite costs, lack of time, and no need for such training for others than those who are to deal with high-risk situations.

We don't train those who need the most training.

And who are they? The "first responders" a catch phrase now used to describe those who arrive on the scene of an accident/disaster/crime in progress, generally just after its occurrence but quite often while it is still going on.

Here is the basic contradiction between mission and training that underlies the whole problem. The first responder, who faces the most danger, gets the least amount of training.

This situation is now being ameliorated somewhat in certain situations. With the realization that our country is in real danger from weapons of mass destruction (loosely defined as chemical, radiological, biological, or nuclear), we have embarked on a crash course of training for first responders. We recognize that the actions they take may, literally, mean the difference between life and death for hundreds or perhaps thousands of people.

But we don't offer specialized training for the people who are first on the scene for the thousands of incidents that take place every day in every part of the country. One could call these incidents involving weapons of individual destruction. The hot burglary gone wrong, where the felon ends up holding the homeowners hostage. The domestic disturbance that escalates into open warfare. The drive-by shootings and home invasions, convenience store holdups, and armed muggings that seem so much a part of present-day society. Never mind that the incidence of violent crime in the United States has dropped precipitously in recent years (largely because we are locking up the violent felons and keeping them there). If you are a patrol officer responding to one of these incidents in progress, the crime rate in your area is exactly 100 percent. And you have to deal with it.

I'll make a flat-out statement here: *every* police officer

needs the tactical training that has so far been provided only for SWAT officers. Every infantryman must be able to fight in specialized situations (e.g., urban terrain) with the same level of expertise so far taught only to special operations units. Every armed security officer must have the ability to respond effectively to potential lethal-force encounters in every area within his purview. NTF, used properly, enables this.

Let's look at some job-specific training examples.

Undercover Cops

Undercover officers, if armed at all, have to conceal their weapons in, for instance, ankle holsters, inside-the-pants holsters, fanny packs, or a coat pocket. The method of concealment can and often does change with the season and climate. Now, this can mean setting oneself up for a fall. Experience has shown that a weapon must be carried in exactly the same place, all the time, otherwise too often the officer will be clawing for it in a place where it isn't.

This in itself is bad enough. What makes it worse is that few, if any, undercover officers go through range drills that take into account the position in which they ordinarily carry the weapon. Range qualification depends on high scores. It is difficult to get high scores when fishing for a weapon that is hard to get to, which probably doesn't have the effective range of the normal service pistol, and which may or may not be well maintained.

Further, the engagements that undercover officers usually face are seldom stand-up, face-the-target, aim-for-the-X-ring situations. For example, ask narcotics officers where they do a lot of their work, and their answer is probably bars and similar establishments. Yet, how many of them have practiced drawing their weapon while sitting on a barstool or, worse, in a booth?

NTF cannot solve many of these problems, but it can facilitate a training program that undercover officers can practice in environments similar to those they're familiar with. The same shoot house in which tactical officers are trained can be reconfigured into a bar, a flophouse, or a methamphetamine lab. Officers can practice drawing their weapons from whatever carry, fire at single or multiple targets at ranges normally encountered, and correct the inevitable failures of training/practice in an environment in which neither they nor innocent bystanders are in danger.

Motorized Officers

Those who have ever had to fire at an adversary while on the move in a motor vehicle hope to never have to do so again. This is true even in a squad car, when one officer is driving while the other shoots, and it is much more the case when there is only one officer who has to shoot and drive at the same time. Multiply that problem by several orders of magnitude for the officer on a big motorcycle, traveling at 90 miles per hour and knowing that bullets are whistling by his head.

Do we train our officers to deal with these situations? In most cases, no. Drive-through ranges are so rare that they elicit amazement when encountered. Here, again, the big problem is in safety. You try to keep your bullets within the authorized impact area as you move in a motorized vehicle, maybe from one side to the other, in trying to engage targets. However, with its vastly decreased maximum range, along with anti-ricochet qualities and lack of toxicity, NTF ammo can be, and is, used in such training.

Another chapter with job-specific training needs—burglary squads, jump-out teams, VIP protection units, and so forth—could be written. But, it is still the case that few, if any, of these units get the training, keyed for their specific duties, that they need.

CHAPTER 10

Nontoxic Frangible Training Devices and Facilities

As mentioned earlier, NTF ammunition can save money in the costs of cleanup and the up-front costs for training devices and facilities. The fact that NTF causes far less damage to the impact zone of targets and backstops allows for some superbly innovative and cost-effective training.

TARGETS

Anyone who has shot against steel targets, and has inspected the average target that has been in use for any length of time, is familiar with the damage that results (i.e., the concave shape caused by repeated strikes in the kill zone). Sooner or later, the target becomes cracked so badly that it has to be replaced.

For the most part, this wear is inevitable because of the

TYPICAL WELL-WORN STEEL TARGETS.

physics of impact for conventional ammunition. Striking dead-on (as we hope most shots do), practically all the energy of the round is expended against the target. Yes, the jacket often flies off, but the lead core, where most of the mass is contained, is flattened against the target and then generally falls to the ground. It's the old story of an irresistible force meeting an immovable object. Something has to give. And while what gives way is mostly the rounds themselves, the repeated impact and energy transfer by those rounds have the inevitable effect.

Frangible ammo, on the other hand, has vastly different energy transfer mechanisms. The initial impact may be as hard as (or even harder than) that of conventional ammunition. However, that impact is quickly transferred into the breakup of the round, the remaining energy disappearing with the tiny fragments of the bullet as they dissipate into a spray 90 degrees in all directions from the impact.

What does this mean in practice? It means that the target can be constructed of far less sturdy (and thus far less costly and heavy) material. More than 10 years of use at our facility has shown that steel sheeting of 1/4-inch thickness suffices admirably for any NTF pistol round. The pistol targets should be slightly harder than mild-steel plate (usually a steel of Brinnell 400 or so in hardness is enough). Targets that have been hit by tens of thousands of rounds show no more effect than a bright copper sheen from the bullet residue.

Faster bullets, such as the .223-caliber (5.56mm NATO), require a steel hardness of at least Brinnell 500, and to prevent wallowing should be of at least 1/4-inch, and preferably 3/8-inch, thickness. This is not necessarily because of harder impact (though the energy transfer of .223 is tremendous), but because of the heat generated by such high-speed impact. The .223 NTF will literally burn its way into, and sometimes through, softer steel.

Still, even with the .223-caliber, the thickness and hard-

Typical NTF targets used at the author's facility.

ness necessary for target steel are far less than those needed for conventional ammunition. This translates into lower cost, greater ease in moving targets and target systems, and increased versatility.

There is one word of caution, however. Do not use targets that have been heavily scarred by conventional ammunition in close-in shooting training. A heavily scarred and wallowed-out target will shoot back at you, even if you are using NTF. True, the fragments are small, but they still possess enough energy to cause flesh wounds if you are standing too close to the target.

NTF targets for close-in shooting should be flat-surfaced and unscarred. Standard steel targets designed for conventional ammunition will work just fine; purpose-built targets designed for NTF are much better. But the usefulness of NTF in designing target systems does not stop here.

SHOOT HOUSES

As discussed earlier, the lack of realistic training facilities is one of the major problems in training. Shoot houses are just too expensive to construct, maintain, and keep safe. Ricochet, shoot-through, and splashback make it far too dangerous to use buildings that are not specifically designed for use with conventional ammunition.

Although one cannot just go into a derelict building and start shooting NTF, the effort involved in modifying such buildings is far less than in modifying them for use with standard ammunition. Several factors must be taken into account, but these factors can be dealt with appropriately.

Shoot-Through

NTF penetrates the materials normally found in wood-frame construction (typically 2 x 4-inch studs covered with drywall), though less than conventional rounds. NTF ordinarily does not penetrate concrete-block construction, particularly if the blocks have been filled with concrete or other materials. The rounds do, however, spall concrete and, if fired repeatedly at one spot, will eventually chip away enough to allow shoot-through. Finally, NTF will certainly penetrate the typical steel building walls, most of which are constructed of fairly lightweight, thin material.

Ricochet

Any bullet, including NTF, will ricochet if fired at an acute enough angle. The difference between NTF and standard ammunition, however, is that the NTF ricochet consists of small pieces of bullet material that lose velocity very quickly, rather than large chunks that bounce around for a long time. The difference is also in angle of attack. While conventional ammunition ricochets when fired at any angle less or more than 90 degrees, NTF ricochets only when fired at angles of

10 degrees or less. What this means in practice is that the shooter can engage targets in a much wider arc when shooting inside a building.

Backsplatter/Splashback

As explained earlier, dangerous backsplatter is almost eliminated when using NTF. The caution for the trainer setting up targets in a shoot house is to be aware that there will be a small amount of sidesplatter, which quickly loses energy and the ability to cause damage but may still produce small flesh wounds in a person standing very close.

Modifying Existing Structures

So, how do we modify a building to enable safe training with NTF?

Obviously, frame or steel buildings are going to require more work than do concrete-block or tilt-up-wall buildings. But the modifications can still be made.

1. *Make sure all utilities are turned off and that gas lines are completely drained.* Direct strikes by NTF rounds can certainly penetrate electrical wires, conduits, thinner gas lines, and plumbing. Even though the strike by an NTF round probably doesn't produce enough heat to ignite natural gas, we have found no one willing to test that theory.
2. *Carefully plan the scenario.* Make sure that target placement is such that a shooter won't fire backward or into zones where other shooters might be. Careful target placement alone is one of the major safety concerns.
3. *Plan for backstops.* If using steel targets, the target itself is the primary backstop. However, it must be expected that the occasional round will miss the target. We recommend that steel plates be erected and secured behind the target areas in frame or steel buildings. For pistol rounds, this

plate need not be of expensive hardened steel. Mild-steel plate of 1/4-inch thickness will stop and break up the rounds, and, unless struck repeatedly in the same zone, will last for many more scenarios than you will run. If the primary target is missed often enough to wallow out the backstop, you have a training problem that is probably not going to be solved by NTF ammo.

Concrete-block buildings, particularly those with filled blocks, can be used without the extra backstop, although the trainer may still want to provide a backstop if misses are frequent. Even though the NTF pistol round will not penetrate the blocks, it will chip away at them until sooner or later they will suffer shoot-through.

4. *Consider the angle of attack possible in the various target engagement scenarios and plan for secondary backstops if necessary.* Consider the illustration below. Straight-on

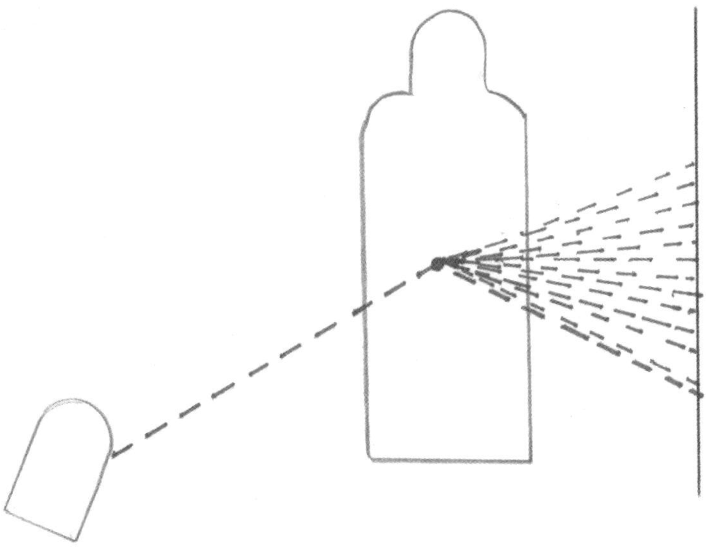

ANTI-RICOCHET PANEL FOR ACUTE-ANGLE SHOOTING.

hits will disintegrate against either a steel target or backstop, as will rounds fired at most angles. It is only when the round is fired at an extremely acute angle that pieces of it will ricochet. Place secondary backstops at a 90-degree angle from the primary backstop so that any chunks of projectile will then strike the secondary backstop, further breaking them into small fragments that quickly lose velocity.

5. *Eliminate weak spots.* If you are lucky enough to find a concrete building that is completely enclosed, you won't have many weak spots. This means areas where projectiles in whole or part may escape the structure. Most buildings, however, are going to have at least a few windows, and of course there will always be doors.

Our first shoot house was a concrete-block building on a shut-down U.S. Air Force base. The building had a number of windows. Our solution was simply to cut 1/4-inch mild steel plates to fit each of these window openings and fix them in place. The total cost for modifying this building, in which we could now engage targets in a full 360-degree circle, was less than $200.

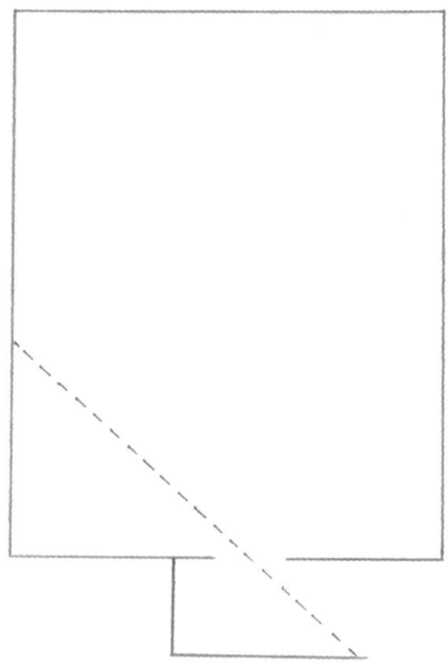

Maximum angle at which rounds can exit a door.

We did not attempt to armor the door, relying instead

on good command and control procedures to eliminate the possibility of the student shooting back through it. If, however, we'd wanted to, we would have erected a simple "L" structure of mild-steel plate (see illustration on page 114) outside the door. Rounds fired into this area from any angle would hit the plate and disintegrate. Maintenance of this facility included frequent inspections to ensure no shoot-through in the concrete and daily sweepings to remove expended brass.

Given these considerations, the possibilities for using existing structures as field-expedient training facilities are obvious. There are few municipalities without derelict buildings, many seized for nonpayment of taxes. Some of these might not be suitable for use (e.g., too much noise, neighbor complaints, too many weak spots), but others will. Look around your own jurisdiction and see what might work.

Purpose-Built Shoot Houses

Perhaps you don't want to use existing structures for your shoot house or cannot because of various restrictions. Don't despair. There are other ways of doing this, and it's cheaper than you might think.

Steel Houses

The illustration on page 116 shows the design of a purpose-built steel shoot house offered by one manufacturer. A word of caution: this house can be used *only* with frangible ammunition. Any other type of ammunition *will* result in trainees' or instructors' facing danger to life and limb.

This shoot house is constructed of 1/4-inch steel plate with a Brinnell hardness of 500, making it suitable for use not only with frangible pistol rounds, but also frangible rifle and shotgun projectiles. The joints at each plate are machined to ensure that there area no gaps. Connectors and braces are provided for stability.

The advantages of such facilities are twofold. First, the

THE BASIC RANGE SHOWN IN TYPICAL EFFICIENCY-APARTMENT CONFIGURATION OFFERED BY ONE MANUFACTURER OF PURPOSE-BUILT SHOOT HOUSES.

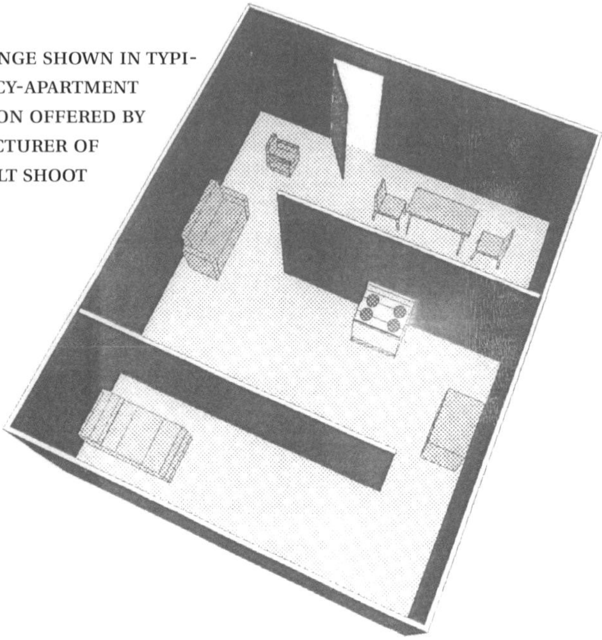

structure can be modified into different configurations. Each plate weighs about 350 pounds, making it hard to move by hand but easy by handcart. The connectors are interchangeable. Modification of the house makes for a better training experience, the trainee being far less likely to get bored with the same old thing, time after time.

Second, it is an easy task to add onto the basic kit, either by individual panel or by expansion modules. Thus, an agency can purchase the basic module and add to it when funds become available. There is virtually no limit to the number and type of panels that can be added to the structure—it even allows for a second story with the appropriate adaptors.

The best part is that the basic structure can be had, on a chosen site and erected, for less than the price of the average patrol car.

Field-Expedient Shoot Houses

Can't afford the steel house? Let's explore other solutions.

The first thing you must have is an area surrounded by walls that can absorb bullet strikes. What type of walls might that be? Dirt is probably the simplest and cheapest medium. With a couple of hours of bulldozer time, berms such as seen in the illustration on page 118 can be thrown up. These berms should be at least 8 feet high, and the entrance/exit should be offset by a barrier as shown.

Within the confines of the berms, construct a shoot house out of the cheapest materials possible. If wind isn't too much of a problem, the easiest method of construction is to erect a PVC framework and stretch plastic or canvas over it. The framework can be as complicated or as simple as you wish. You can use any kind of target inside (though steel is still the best) and be confident that the frangible rounds shot will not ricochet or skip over the berm, and those hitting into it will not create an environmental hazard.

Obviously, good command and control is paramount. No more than one team can use the facility at one time, and that team must be carefully supervised to make sure that they stay together, with every team member in every other member's line of sight at all times.

I know of other agencies that have built temporary facilities in rock quarries, whose sides rise well above 8 feet all around. Standard ammunition could not be used safely in such a place, but frangible ammo can.

Other agencies have used natural terrain features, such as nearby hillsides. For any such training facility, the only requirement is some bullet-absorbing barrier all around the structure.

The only limitation is your imagination. Look around, see what you have. Then imagine what you can do with it, given a good frangible ammunition. You should find that your training program can be immeasurably enriched.

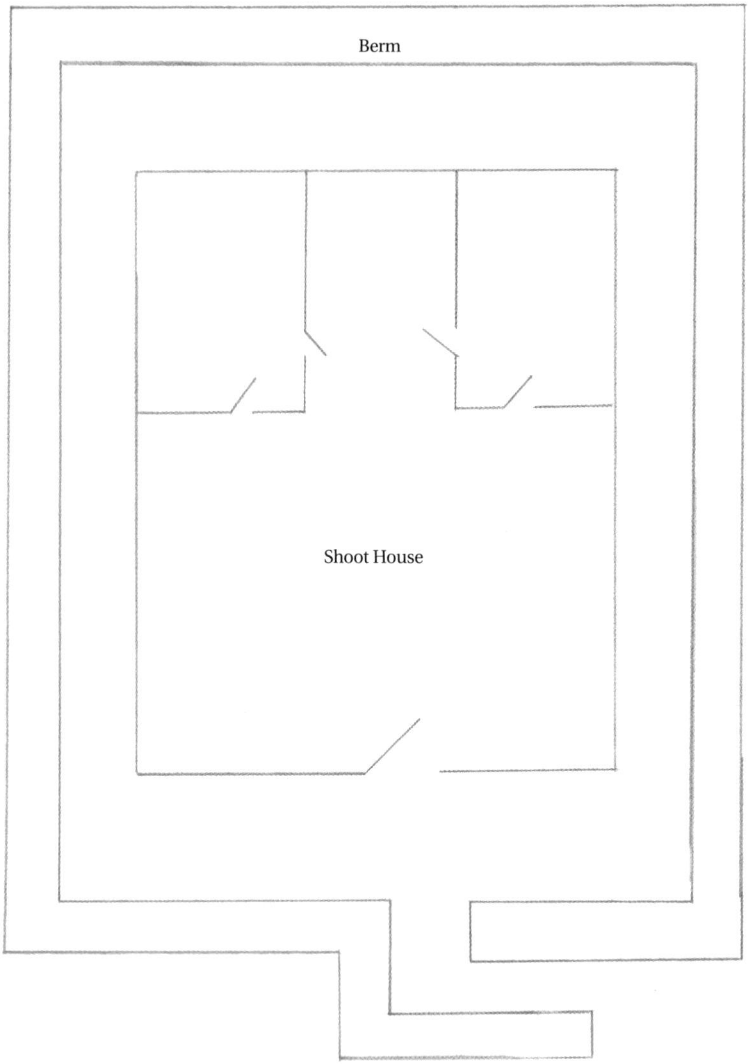

If your shoot house has no natural barrier around it, consider putting a berm around it. Remember that the berm should be at least 8 feet high to contain any expended rounds that leave the structure.

CHAPTER 11

Costs

With all its advantages, why aren't more people using frangible ammunition? Easy answer. It costs more.

In a day where 9mm ammunition from former Soviet bloc countries sells for $120–$140/thousand retail, and even range ammunition made in the United States is not a great deal more, resistance to purchasing an ammunition that costs well over $200/thousand is great. Police departments, in particular, seldom have all the money they need. Convincing the purchasing officer to spend more for the same amount of ammunition is tough.

However, comparing only the purchase prices of ammunition while making the decision leaves out a number of important factors. Purchase price is by no means the only cost.

DOWNSTREAM COSTS

Environmental Cleanup

The U.S. Army, quoted in *National Defense Magazine*, estimates that the downstream costs for environmental cleanup will add at least 30 percent to the cost of leaded ammunition. This figure has been judged conservative, given the huge efforts undertaken at military bases that have been closed. For instance, at Fort Ord, California, the firing ranges were often situated on the beaches of Monterey Bay, and their fans extended well out into the water. Not only does the Department of Defense (DOD) now have to remove existing berms, treating everything in them as hazmat, it must also sift all the sand in front of and out past the berms to remove all the rounds that missed the berm. There is even a move afoot to force DOD to clean up the ocean floor out past the beach. One can only imagine what that will cost.

Before you say, well, that's them, and we don't have that kind of problem, take a look at your own range. Does the water from the range flow into any natural watershed? Even ravines and ordinarily nonflooding areas count. Now imagine what it would take to clean up everything downstream for the couple of miles that spent rounds may have washed to.

And since most of the military's ranges are outdoors, the 30-percent figure does not take into full account the costs of maintaining and cleaning up an indoor range. Included in these figures are the replacement and disposal of filters, removal of spent rounds, health checks for instructors and range personnel, laundry, cleanup, and various other items.

For sake of argument, let's stay with the 30-percent figure. Now the round that may have cost $150 per thousand has an add-on of $45, for a total cost of $195 per thousand.

Hidden Costs

How many workmen's compensation claims has your agency paid for wounds from ricochet and splashback? If none, yours is either a small agency or very lucky. Consider the case of an officer in Florida, who was hit by a 12-gauge slug that splashed back almost whole from a distance of 25 meters. Luckily for him it hit his lowest rib, traveling around the rib and laying open skin, muscle, nerves, and blood vessels down to the bone, almost to his back. I say "luckily for him," because a fraction of an inch lower and it would have penetrated just beneath the ribcage on the right side—where the liver is.

In another case, an officer was spotting for a rifleman shooting at steel targets, again at 25 meters. A piece of the jacket came back just as the officer lowered his head, passing over the top of his protective lens and lodging in his eyeball. This officer was lucky as well: he didn't lose his eyesight.

Each year hundreds of workmen's comp claims come into the state offices, attached to similar stories. Many people suffer only temporary disability; others are out for life. What does it cost to pay an officer full disability for the remainder of his life?

But it doesn't stop there. Increasingly we are seeing workmen's comp claims for lead poisoning. Lead toxicity is a terribly debilitating disease. What are the costs to the agency for such claims? Worse, what are the costs to the family and society as a whole?

It gets worse. Lead toxicity is particularly pernicious in its effect on a fetus. How many females of childbearing age do you have in your agency? Want to bet what the jury will say when one of them comes up with a child who is born with the problems that lead can cause? Want to guess what even a mediocre lawyer can get out of that?

We've discussed the problems many agencies have in finding a range on which to train. With firing ranges being closed down left and right, even large agencies find themselves having, at best, to share a range and, at worst, having to travel long distances before getting to the shooting facility.

A natural tendency is, then, to shave the amount of training given to the officers to the bare minimum required by state or city ordnance. Good solution all around: less cost for ammo, more street time for the officers, fewer instructor hours needed.

This is fine until a shooting takes place. It doesn't even have to be a bad shoot. In today's climate, any shooting is going to be questioned. At best, the questioning comes from inside the agency; worse, by the baying hounds of the media; and at the absolute worst, in a courtroom where there is a lawyer determined to get far more than a pound of flesh. *All* training is going to be questioned. And if the plaintiff's side finds out during discovery (and it will) that firearms training has been cut back—even if it still meets state, city, and federal mandates—the plaintiff has a case. At that point, smart officials are going to start talking to the city attorneys, telling them to find a way to settle.

Finally, what is the cost of an officer killed or wounded in the line of duty because of improper or insufficient training?

Obviously I cannot come up with a figure, as we did in the environmental section, to add on to the cost of the bullet, given such eventualities. But I think I can safely say that there is a cost, and it can be staggering.

Bang for the Buck

Even without taking into account environmental costs and hidden costs, there is one simple fact that should make you consider the use of NTF ammunition. You cannot do the same things with conventional ammunition that you can with NTF. If nothing else, what you have read so far should convince you that you can enhance and enrich your training program by using NTF.

Low-cost shoot houses; up-close engagement of steel targets; increased use of existing range facilities; reopening, in many cases, ranges that might have been closed; and increased officer safety are but a few things that give you a lot more bang for the bucks you spend.

The question becomes not whether *you can afford to*, but rather *whether you can afford not to*.

CHAPTER 12

How to Evaluate Frangible Ammunition

Okay, you've decided you might want to try out some frangible ammunition. Now you look for sources and find that there are many. Each has competing claims. Each has a reason that it is the best, and if you try the other guy's stuff, well, you're going to be disappointed, at best, or disgusted, at worst.

Should you change your duty ammunition without testing the new stuff? Of course not. Neither should you buy a batch of frangible without first trying it and all the others. Some of the tests will be familiar, being the same ones you'd use for standard ammunition. Others are not. That's where we come in.

ACCURACY

The only proper way to test for handgun accuracy is on a Ransom Rest. And the only test that has any validity is by firing the weapon you are going to use daily. Some government agencies insist on testing for accuracy using test barrels. The fact is, though, that most of these test barrels do not necessarily replicate the characteristics of the service pistol (e.g., different twists, lockup mechanism not replicated, different barrel lengths, different bore tolerances). You are not going to shoot the ammunition through test barrels when you get to the range. Why should you test for accuracy with them?

Make sure the Ransom Rest is installed according to the manufacturer's specifications. Lock the pistol in place and fire three rounds of standard ammunition to settle the pistol. Retighten the rest; fire three more rounds. Check for any wild rounds, which may indicate instability in the mount.

Now you are ready for grouping. Load six rounds of your duty-carry ammunition into the magazine and slow-fire at the target. Ensure that the weapon is clear and then go downrange and measure the target. Group size should be measured from the rounds farthest apart, generally to the center of the hole. Note the center of mass of the impact zone. Replace the target. Return to the test table, load six rounds of frangible, and repeat the exercise.

Some frangible ammo manufacturers will protest that because frangible is intended for short-range use against steel targets, it does not have to be as accurate as the service load. Nonsense. There is no physical reason for frangible to be any less accurate than service ammunition—in fact, by its very nature (monobody construction), it can often be *more* accurate than service ammo.

In the same manner, test all the frangible ammunition you have accumulated. Set your own standards with regard to accuracy (i.e., how important accuracy is as opposed to frangibility or weapons function, for example). Give each brand a score according to these standards.

By the way, the frangible rounds should have a center-of-mass group not much different from that of your service ammunition. You don't want to have to change the sights on the weapon just to shoot frangible.

Rifle ammunition can be tested in the same manner, albeit without the benefit of a Ransom Rest. The rifle should be supported by sandbags or rests, with extreme care taken to ensure the same sight picture each time. I find that the use of telescopic sights helps in this. Rifle ammunition should be fired at realistic distances—you are not going to use frangible as a long-range sniper round, so why test it that way?

FUNCTION

You want your frangible ammunition to function at least as well as your service ammunition. Frangible ammunition, because of its composition, is always going to be lighter than leaded ammunition. Lead is quite dense, far more dense than almost any other metal practicable for use in ammunition. Bullets, by their very nature, have to fit in a certain physical space. Therefore, the space filled by lead is going to weigh more than the space filled with some other material. This can create function problems in semiautomatic weapons if the bullet is not properly designed and loaded.

Semiautomatics can suffer function problems even with standard lead ammunition if they're not used properly. Firearms instructors refer to the problem as "limp-wristing" or "riding the recoil." Semiautos depend on the recoil and backpressure from the fired round to move the slide back to the rear, where it picks up and chambers another round. They also depend on the firer maintaining a stable firing platform. When the shooter holds the weapon in a limp wrist, or when he allows the weapon to ride back upon firing, this stable platform is not provided. The weapon will often not fully extract the spent cartridge, will not pick up the next round, or will "stovepipe" the spent round.

We simulate the limp-wristed shooter when we test frangible ammunition, holding the weapon in the loosest possible grip and attempting to get it to misfire. You should perform the same tests. Pick a group of shooters, some of whom should be of slight stature, and instruct them to deliberately hold the weapon loosely. Give them a magazine loaded with 10 rounds, not allowing them to see what type of ammunition is in the magazine (this helps eliminate subconscious prejudice). Fire 10 rounds of your normal service ammunition. Note any misfires. Now fire 10 rounds of the frangible you are testing. There should be no more misfires than with the service ammunition. By misfires, we mean the occurrence of stovepiping, nonextraction of the spent round, double feeds, failure to pick up the next round, and the slide's not locking back after the last round.

A properly designed and loaded NTF ammunition should result in no more misfires than does the service ammunition with which you arm your people.

RELIABILITY

I'll say this at the outset: present state-of-the-art lead-free primers have reliability problems. The chemical used in these primers, dinitrophenol, is simply not as sensitive as the lead styphenate commonly used in standard primers. Even though this problem is being worked on by manufacturers all over the world, it has not yet been solved.

How bad is this problem? It often depends on the type of weapon being used. The .38 Special seems to suffer the worst. If the firing pin or hammer is worn a little bit, as it often is in a service revolver, the problem becomes worse.

Your job, then, is to test the frangible round in a selection of the service pistols you are using. Note the number of misfires in, say, a thousand rounds. There may be none at all. There may be as many as 10 (more than 10 misfires and the lot should probably be rejected, there being more problems than just primer sensitivity). Ask yourself, if you can live with this number of misfires while training. After all, any weapon can misfire, and firearms instructors often deliberately induce a misfire as a part of the training process. Obviously, you don't want too many misfires, otherwise valuable training time will be lost on misfire drills when it should be spent on accuracy and tactics drills. You have to make the decision as to how many is too many.

This reliability problem is why we do not recommend NTF, in its present state of the art, for service use. Even one misfire is way too many when the one round may mean the difference between life and death.

For those people who do wish to use frangible as a service round (and service applications are discussed in the next chapter), I recommend that the frangible be loaded with standard primers. Yes, this creates toxicity, the lead-styphenate primer spewing lead particles every time a round is fired. However, if the weapon has to be used in a lethal-force incident, the lead particles in the air are probably the least of the worry. Some users now routinely train with frangible and nontoxic primers, and change to frangible with leaded primers for service use. There is no measurable difference in performance between the two, except for the occasional misfire with the lead-free primer.

FRANGIBILITY

How frangible is frangible? What makes a frangible round? How do we define frangible, anyway?

If only the dictionary definition—"breakable or brittle"—is considered, many kinds of ammunition could be classified as frangible. Bullets consisting of a copper jacket loaded with steel or lead shot, intended to break up and create larger wound channels in flesh, for instance, are technically frangible. But it would not be a good idea to shoot them at close range at steel targets.

And therein lies the method of determining frangibility. What do we want the frangible ammunition to do? If we judge it by its characteristics and function, then we can devise tests to make sure it meets those functions.

For the purposes of this book, I will speak of frangible ammunition that is designed and marketed to lessen the hazard of ricochet and splashback. You want to be able to engage hard targets at combat ranges, without producing excessive danger to the shooter or other personnel.

It is not enough to depend on the manufacturer's claims. Each thinks its own products superior to any of the others and will quickly tell you about the disadvantages of the competition's and the merits of its own.

I suggest that the only responsible course of action is to perform the tests on your own, there being different circumstances in almost each location, training program, and group of shooters. What might work very well if shot against a vertical steel plate at a 90-degree angle could offer difficulties if shot at plates placed at another angle, if shot at movable plates, or if subject to any of a host of other factors.

Various tests have been devised to measure this quality of frangibility. One federal agency has constructed a steel box into which the rounds are fired. The round strikes on the angled steel plate (the angle of which can be changed to simulate almost any possible firing surface), and the fragments of the round are caught in the water bath beneath the impact plate.

This is a good method, but it depends on the criteria set up for judging the results. For instance, some agencies believe that it all depends on the size of the fragments left after impact. X numbers of fragments of Y size or larger and the round is noncompliant; Z numbers of fragments of W to Y size, and the round is marginal, and so forth.

The only problem with this approach is that it does not take into account the nature of the fragments themselves. All other things being equal, a relatively large, rounded fragment is going to cause less damage to human tissue than a smaller, sharper fragment. Look at the incidents of splashback injury in your agency. Have they been caused by the lead core or the jacket? In most cases, it's the jacket.

A simpler test is the penetration test. After all, what you are trying to do is to eliminate injuries to your shooters and bystanders. To injure, the fragment must penetrate; bruises don't count.

Set up your steel backstops at all angles at which they may be engaged by the shooter. Place 1/4-inch cardboard witness panels at all locations where the shooter or any bystanders might be. Fire the round and note all penetrations of the witness panels. Even one penetration is too many when you consider that penetration might be in someone's unprotected eye. Fragments lodged in the surface of the cardboard can largely be discounted because human skin is a remarkably resilient medium. The fragments that might lodge in the relatively soft cardboard are going to bounce off skin.

The only problem found thus far with this method is when testers place the cardboard witness panels at unrealistic distances or angles from the impact plate. Remember, the fragments of the round must go somewhere—they do not simply disappear. Place a witness panel within a few inches of the impact point at a 90-degree angle from the surface of the plate, and you are going to get penetration from any frangible round on the market. One hopes that your range controls will keep bystanders from standing right next to the target, where they most certainly will be struck by the spray of even the best designed and manufactured frangible round.

I cannot stress enough the importance of this testing. You are working to lessen the danger to your people—you must test at every conceivable angle and situation. Only then can you be sure the claims of the manufacturer are true and will work for you.

PRESSURE

Pressure can only be tested by using test barrels and appropriate measuring devices (either copper-crusher or piezoelectric transducers). Acceptable pressures are set by the Sporting Arms and Ammunition Manufacturer's Institute (SAAMI) and, for the military, NATO. These pressures should never be exceeded. Otherwise you stand an excellent chance of causing damage to the weapon and injury to the shooter and bystanders. Those handloaders who have exceeded pressures and suffered case-wall failure can tell you what an unpleasant experience it is.

Reliable manufacturers are going to certify that chamber pressures are within SAAMI standards. The problem? There are, as yet, no standards for frangible ammunition. Still, pressures should be well within those for standard ammo.

You should still check for signs that pressures may be exceeding those considered safe. A simple method is to check expended brass. Excessive flattening or cratering of the primer are signs that pressures are high, though they may well still be within standard. Worse is bulging of the case wall. The weapons that will show this the most clearly are those with portions of the chambers unsupported (such as the Glock). Excessive pressures will result in a noticeable bulge at this location, and the ammunition should immediately be taken off the firing line. And before anyone asks why there should be a weapon with a portion of the chamber unsupported, consider the purpose. Should, God forbid, someone shoot a weapon that has a barrel obstruction, the blowout of the case wall will be directed downward, usually blowing out the magazine and often damaging the magazine catch—

which is a much better than having the slide blown back through your head.

If you have reason to believe chamber pressures are being exceeded, you should have the ammunition tested. If your local agency does not have that capability, there are a number of nationally recognized labs that will do the job for a fee.

BARREL WEAR

Why do we include barrel wear as a category for testing? Because barrels are designed for shooting with standard ammunition—either lead or copper—or other soft-metal-clad lead. The hardened steel in the barrel is designed to hold up under such use, at least for a reasonable number of rounds. Everyone has heard of shot-out barrels, some may even have experienced them. Even with the relatively soft metals used in standard bullets, shoot enough of them and they will inevitably wear away at the steel.

How much worse, then, is it to shoot something that is harder or more abrasive? Look at the Materials Safety Data Sheets (MSDS) supplied with your frangible ammunition. What metals are in the projectile? If they are anything other than relatively soft material (e.g., copper, zinc, tin, bronze, brass), you may have a problem.

Some manufacturers include harder materials because of superior density. They then clad the bullet in a copper jacket to reduce the possibility of abrasion from the harder material. Whether this solution creates problems with frangibility, only you can judge.

Others claim that the polymer binder so completely coats the material that abrasion is reduced or eliminated. Anecdotal evidence from the field casts doubt on this claim. Again, only you can be the judge. If you are using a bullet that contains harder and more abrasive materials, it may be useful to set up a test barrel. Gauge the new barrel with precision instruments (many barrel makers will air-gauge it for you). Fire set amounts of the ammunition through the barrel and regauge it. If there is wear beyond the amount considered realistic by the barrel or weapons manufacturer, you have a problem. Continued shooting of this type of ammunition is inevitably going to reduce the accuracy of your weapons.

After having done all these tests, and having found a frangible ammunition that meets your criteria, you can feel fairly safe in using it for training. The next question is, can it be used on duty?

As in so many things, the answer is complicated. Only you can make the decision. Chapter 13 will help you make it.

CHAPTER 13

Nontoxic Frangible Ammunition on the Street?

The question that inevitably arises is this: Can NTF be used as a service round? The answer is a qualified yes. A very qualified yes. A number of arguments, both pro and con, can be advanced.

But first a couple of issues need to be dealt with. I'll say at the outset that there is *no* perfect service round, despite claims to the contrary.

What would a perfect service round do?

- It should give you one-shot stops, every time, without fail.
- Performance should not be affected by clothing, body type, or light barriers (e.g., automobile glass, sheet metal, Sheetrock).
- It should never, ever overpenetrate.

- If the round misses its target, it should offer no danger to innocent bystanders or other officers.
- Recoil should be such that even officers of light build are able to easily absorb it and come back on target within a fraction of a second.
- The entire round should be contained in a package small enough that magazines, and thus magazine wells and handgrips, are small enough to accommodate the grip of people with small hands.

Can you think of such a round? Neither can I, and I've been in this business nearly 40 years. Quite simply, it doesn't exist.

Keep this fact in mind when evaluating NTF for service use. Neither is NTF, in any configuration, the perfect service round. But it can offer certain advantages in specific situations.

GENERAL CONSIDERATIONS—PRO

Lethality

Actually, lethality could be listed in both the pro and con sections. Frangible pistol ammunition designed for training is, quite simply, not lethal enough for service use. Each round we have tested acts much the same as ball ammunition; that is, no expansion, no tumbling, and no fragmentation *even when it encounters a major bone.*

Such information may come as a surprise to those who, without benefit of actual testing, think that frangible ammunition will, as I have heard said, "simply annihilate whomever it hits." Think about it. Frangible ammunition is designed to disintegrate only *when it strikes something as hard as or harder than itself.* Human or animal tissue hardly meets that criterion. And those familiar with anatomy and physiology know that bones contained in live flesh are nowhere near as hard as bones that have been taken out and dried.

Instead of fragmenting, the rounds we have tested have

actually *overpenetrated* the media into which they've been shot. Pistol rounds, according to FBI standards, should be able to penetrate from 12 to 14 inches of 10-percent ballistic gelatin, going no farther than 14 inches in order to avoid overpenetration on the average human being. The frangible rounds we've tested have penetrated, on the average, 18 to 26 inches of gelatin. We then conducted the tests using various beef bones still encased in flesh to simulate common body parts and found no major difference in performance. We then shot the rounds into pig heads obtained from a local butcher. High-speed photography shows the round exiting the back of the skull still whole.

When you think about it, you realize that this is not an undesirable trait in training. You don't want to train with a round—should there be an unfortunate accident—that can cause damage to other trainees or instructors. Being hit with a round that passes clear through, leaving an entrance and exit hole of the same size, is far preferable to being hit by one that expands, fragments, or otherwise creates excessive damage to flesh and bone.

However, we are only talking about frangible pistol rounds designed for training. Frangible pistol rounds specifically modified to be used in service offer vastly different characteristics.

The photograph at the top of the next page shows a block of ballistic gelatin 4 x 4 x 24 inches. A 9mm round fired from 10 meters away completely goes through the block, exiting whole from the rear of the block with enough force to remain frangible when it hit the steel backstop. Entrance and exit holes are approximately the same size. Swirl marks in the gelatin indicate possible secondary wound cavity effects; however, the primary cavity is the same size the entire way through. Other rounds fired into the same block of gelatin can be seen embedded at 20.0 inches and 21.5 inches, respectively.

The photograph on page 141 shows the same round,

A standard NTF bullet shot against 10-percent gelatin and beef ribs simulating a human torso. Note the 20- and 21 1/2-inch depth of penetration.

modified by prefragmentation techniques pioneered by the author. The round, upon impact, fragments into several large chunks, the wound paths diverging almost immediately. Depth of penetration of the deepest chunk is 12 inches. No piece of the projectile exits the ballistic gelatin. Compare this to a service round designed to expand upon impact. Which do you think offers the most chance of hitting something vital?

This characteristic is exhibited no matter what pistol caliber is used (rifle calibers are dealt with later) and no matter what body structure is affected. We performed the same tests as with unprefragmented rounds (i.e., against beef bones encased in ballistic gelatin), finding that the chunks of projectile went through bone and flesh with equal ease. Prefragmented rounds fired against pig skulls entered the skull but did not exit. Examination of the skull content showed most of the pieces of projectile lodged against the

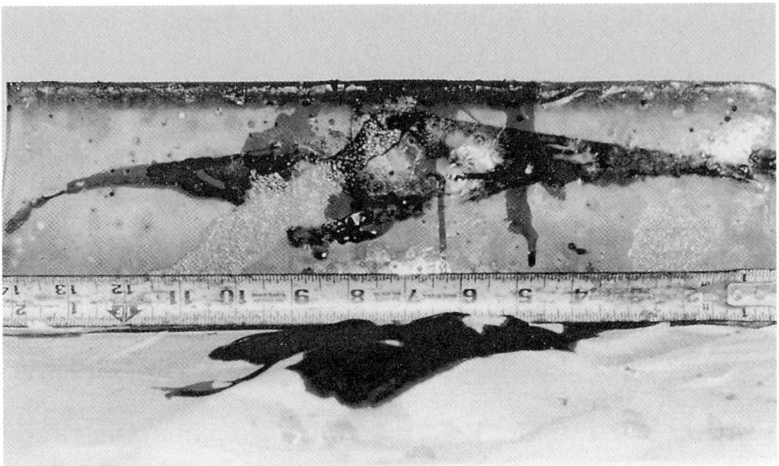

HOLLOWPOINT PENETRATION TESTS: 9MM PARABELLUM (FROM 7 METERS INTO 10-PERCENT BALLISTIC GELATIN).

rear wall of the skull. I leave to your imagination what they did to the contents of the skull.

So, can frangible rounds of the proper design exhibit the necessary lethality? The answer is an unqualified yes. An added advantage is that the modifications to the round are so slight that they change ballistics and performance not at all. Thus a training round can be used that, should the worst happen, offers reduced damage to the recipient of the misdirected bullet, while a round of the same performance and characteristics, but of increased lethality, can be used for lethal-force encounters.

NTF rifle ammunition must be dealt with separately. The vastly increased speed of the round causes NTF rifle ammunition to exhibit characteristics completely different from those of NTF pistol ammo.

This is particularly true of .223-caliber (5.56mm). With the speeds at which this round travels, *everything* is a hard object.

The .223, fired against unshielded ballistic gelatin, begins to come apart upon contact with the medium, fragmenting into a few large pieces and countless small ones. The primary wound channel is around 2 inches in diameter, and secondary expansion of the gelatin is such that the 4-inch-square block will blow a hole in the 3/4-inch piece of plywood that it rests on (see photograph on page 143). Every bit of the energy this round carries (966 foot-pounds) is expended into the gelatin. At very close ranges (7 to 10 meters) the effect is the most pronounced. Out at 100 meters the fragmentation effect is somewhat alleviated, but the round still contains enough energy to penetrate 14 inches of gelatin.

Our experimentation then included simulated body structures, as we had done with the pistol round. Again, beef bones served as rib structures and large-bone impact zones. The round paid little attention to the bone, penetrating about the same amount of gelatin as it did during the pure-gelatin tests. The only difference was that pieces of the bone became secondary projectiles.

We then fired the round through a level-IIIA vest panel, at first without the ballistic shield and then with the ballistic shield in place. The projectile went through the vest, seemingly little slowed, and still penetrated both gelatin and bone structure to a depth of 11 inches. Having the ballistic shield in place made little difference; the round still offered about the same amount of damage to the flesh behind it.

Further tests would have had to be done at ranges past 100 meters, but given what occurred on the test range, I believe that the .223 round will quickly begin to offer no more damage to flesh than does .223 ball, the loss of velocity causing the NTF .223 to remain whole upon impact.

Limited testing was done with .308 Winchester (7.62 NATO) and .30-30 Winchester because there is only one manufacturer that currently markets these rounds. When my grandson and I took the rounds for hunting during deer sea-

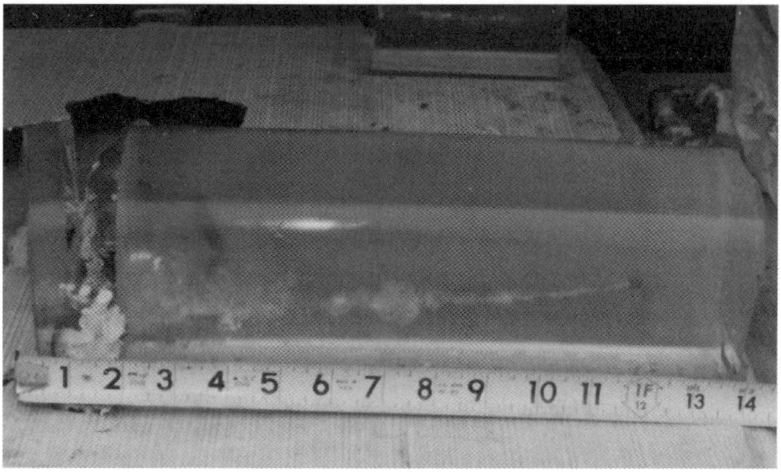

THE EFFECTS OF A .223 SHOT THROUGH A VEST, GELATIN AND THEN RIBS, AND THROUGH GELATIN AGAIN. THIS PHOTOGRAPH SHOWS THE DEPTH OF PENETRATION.

son, I used the .308 to take a whitetail at a distance of 175 yards with a shot that penetrated the shoulder and broke up while doing so. The fragments shredded the heart and liver and came to rest beneath the skin on the opposite side. My grandson, using the .30-30, shot a whitetail at a distance of 50 yards, the round severing the spine of the animal, which was then finished off with a shot to the head.

I believe, then, that the lethality of the NTF round is adequate to the purpose. Will it give one-shot stops under any circumstance? Of course not. But then neither will any other round, despite claims to the contrary.

Penetration versus Overpenetration

As stated in the previous section, NTF pistol ammunition unmodified for service use has a tendency to overpenetrate. Some of the types tested here achieved as much as 26 inches of penetration in 10-percent ballistic gelatin. They will penetrate even farther than standard ball ammunition because of the speed at which they are fired. The lighter round generally is pushed out at 200 to 300 fps faster than standard ammunition.

NTF modified for service use, however, simply does not overpenetrate. The fragments into which the round breaks take vastly different paths through the ballistic media, slowing quickly. Depth of penetration can actually be manipulated, given a slight modification to the round.

I do not have to lecture those who risk confrontations that involve lethal force about what an advantage this is. No overpenetration means vastly reducing the risk of the round coming to rest somewhere other than where it is supposed to. And that leads directly to the next consideration.

Reduced Danger of Collateral Damage

According to statistics released by the U.S. Department of Justice, some 82 to 83 percent of police bullets miss their intended targets, year after year. One hopes that given increased training and the use of more realistic, stress-filled exercises, this percentage can be lowered. However, there will always be the round that goes where it is not intended to go.

Given that most police shootings take place in urban terrain, we have a situation ready-made for tragedy. The standard police bullet is going to ricochet off concrete, stone, steel, some glass—in short, any surface likely in urban terrain. Where does this bullet come to rest? With luck, it expends all its energy harmlessly. Otherwise, it may end up in the body of some bystander or other officer. Incidents of accidental shootings are far too numerous to list here. Suffice it to say, it is a continuing and troublesome problem. Now, multiply the problem by many orders of magnitude if the shooting takes place, say, in a crowded building.

Although a direct-on impact to a bystander with the use of NTF service rounds will still kill or seriously wound the innocent person, the possibility of impact from ricochet can be vastly reduced. Where the standard round will skip, losing only a tiny fraction of its energy, the NTF round will disintegrate. Bystanders off at a certain angle might be injured from bullet spray from the particles but will suffer little more than flesh wounds. At most angles, they will be safe from any harm at all.

And, of course, it is not only the bystanders we worry about. Other officers are just as subject to injury or death from missed rounds. Bullets are notoriously indiscriminate. As a friend of mine once said, “It’s not the bullet that has my name on it that I worry about. It’s the one labeled, ‘To whom it may concern.’”

Retained Energy

Finally, a word about energy. We have no intention here of taking part in the ongoing argument over which is better—light rounds moving faster or heavy rounds moving slower. Both sides have their points, often arguing the same point from different angles.

What there should be no argument about, however, is that retained energy is an important criterion. It is the transfer of energy, as well as the contact between bullet and critical body structures, that makes a difference when considering bullet effectiveness.

We have already shown that the frangible projectile can be modified enough to provide sufficient contact between bullet fragments and critical body structures. But does the NTF round maintain enough energy at combat ranges to be effective?

Consider the following table on page 140.

Table 1
9mm Parabellum
(from Sierra Ballistic Tables)

Given that retained energy is somewhat higher than standard ammunition practically out to the maximum combat range for the round, we feel it can safely be said that no degeneration of performance will be suffered when using NTF in a service application.

Cartridge	Longbow NTF	Winchester Talon	Federal 124 Hydro-Shok
Muzzle velocity (fps)	1,450	990	1,120
Muzzle energy (Foot-pounds)	443	320	345
10-yard velocity	1,365	980	1,098
10-yard energy	393	314	332
30-yard velocity	1,236	962	1,060
30-yard energy	322	302	309
50-yard velocity	1,143	945	1,027
50-yard energy	276	292	290
Maximum range (yards)	1,326	2,281	1,935

GENERAL CONSIDERATIONS—CON

Barrier Penetration

NTF ammo does not do well at all against targets that are shielded by even relatively flimsy barriers. The round will penetrate the barrier—its retained energy pretty well ensures that—but will start to come apart upon contact with the barrier. Targets placed behind barriers such as automobile windshields show a spray of bullet fragments and pieces of windshield embedded in the surface of the gelatin behind it, but it would only be by chance that one of these fragments would retain enough energy to provide a stopping wound.

This is true of Sheetrock, light plywood, the relatively light metal found in car doors, and just about any type of glass.

Of course, this quality can also have positive effects. The armed forces use it to their advantage during military operations in urban terrain (MOUT). Command and control during such operations is generally a nightmare—danger of fratricide from rounds fired through building structures by a friendly unit is high. Using NTF, particularly in .223, vastly decreases this problem. When it hits even light building structures, the .223 comes apart very quickly, offering a danger only to a person leaning against the wall on the other side—and not even then if the individual is wearing a PASGT vest.

I'll leave it up to you do decide whether this characteristic is an advantage or disadvantage, given your particular circumstances.

Firearms Examiner's Bane?

There have been periodic outcries from forensic types, worrying about getting ballistics off a round that disintegrates upon impact. The perfect assassination round, someone once called it.

As with so many things, the reality is somewhat different than the impression. Frangible ammunition maintains barrel marks very well. We have numerous spent bullets and bullet fragments from our experiments and have submitted the fragments to forensic experts for analysis. The general consensus is that, if the round hits flesh, enough of it can be recovered to identify from which weapon it was fired.

Obviously, this is not true of rounds that miss their target, hitting instead a hard surface. The particles are generally far too small to identify.

This is also the case with .223-caliber rounds fired at close range. The .223, as stated earlier, starts coming apart upon contact with almost any surface, fragmenting into particles that are too small to recognize. Of course, this does nothing for the brass, which can be identified as quickly as can the brass from any other round.

Will it make the firearms examiner's job more difficult? Possibly. Is it the "perfect round for assassination"? Hardly. There is just too much chance of leaving something, whether it be a bullet piece or brass or both, that could identify the weapon.

Besides, if an individual wants to confound a firearms examiner, he uses a shotgun.

Lead-Free Primers

The following issue was covered earlier, but it bears repeating here. If you wish to use frangible ammo as a service round, it should be loaded with a conventional leaded primer. This situation may change with the research and development currently taking place in the field, but for now, opt for reliability.

OVERCOMING THE NEGATIVES

As said at the beginning of the chapter, no round is perfect for all service applications. Nor do I think that any such round is coming down the pike in the foreseeable future.

The decision to use frangible as a service round must take into consideration your particular tactical situation. Are you likely to have to fire through barriers (as is the case for road cops)? Can you train your officers to fire multiple shots into the same small area? If so, frangible may still be used against side windows, the first shot shattering the window; follow-on shots taking out the target. Should you be shooting through the windshield of an automobile in the first place? Seems to me that standing in front of a two-ton weapon is a poor tactical choice, but it may be inevitable under some circumstances. Another question to ask yourself is, whether the service round now used offers any particular advantage when fired through a windshield? Many agencies take the word of the manufacturer, only to find the whiz-bang round they have been sold skips off the slanted surface of the glass.

The solution? Analyze your lethal-force incidents. See what worked and what didn't. Determine whether the rounds used in the incident did their job adequately, or whether they created as many problems as they solved. Set up tests that re-create these situations. Test everything.

CHAPTER 14

Nontoxic Frangible Ammunition Environments

Frangible can be used as a service round, but there are some areas in which its application is far more appropriate than in others. Generally, these can be broken down into "hard" and "sensitive" environments, crowded situations, or a combination thereof.

HARD ENVIRONMENTS

For the purposes of this discussion, we'll classify hard environments as those where a majority of the surfaces likely to be encountered by the stray round will probably cause it to ricochet. Practically any urban terrain fits this bill, but some places are harder than others.

THE PAPER THICKNESS IN THIS TEST WAS MEANT TO SIMULATE THE PADDING PRISON INMATES WOULD USE TO PROTECT THEMSELVES DURING A PRISON RIOT.

Corrections Facilities

A corrections facility is a perfect example of a hard environment. Every surface is stone, concrete, steel, or hardened glass, with the possible exception of the exercise yard—and often even that is concrete surfaced. Corrections officers have told us of three-time-loser prisoners who decide to murder or seriously injure another prisoner (or even a guard), surrounding him in the exercise yard and stabbing him repeatedly, in full sight of the tower guards. They take advantage of the fact that the guards are hesitant to shoot, knowing that any round that misses its intended target is likely to ricochet and hit someone it is not supposed to (with dire consequences for the guard). For habitual criminals in prison for life, with no chance of parole anyway, additions to their sentences mean nothing, and without the consequence

THIS PHOTOGRAPH ILLUSTRATES THE PENETRATION OF A .223 ROUND IN GELATIN AFTER GOING THROUGH A JACKET AND NEWSPAPER.

of immediate lethal force restraining them, they feel free to operate with impunity.

Still other corrections officers have told us of situations inside the cell blocks where lethal force would have been appropriate (e.g., hostage situations, riots) but could not be applied because of the probability of collateral damage. For a time, some agencies tried to use specialty rounds that contained buckshot or small fragmented rounds in a copper jacket, only to find out that missed rounds still ricocheted and splashed back, often injuring other officers and even the shooter.

Still another problem described is when prisoners are "getting ready to go to war." They then often dress themselves in makeshift body armor, consisting of layers of newspaper or magazines wrapped around their torsos and secured in place by blankets or jackets. The previously available specialty rounds did little good against such protection, expending all their energy within the first 3 to 4 inches of "armor." Less-than-lethal rounds (e.g., beanbags, baton rounds) were of even less use.

We have experimented with frangible rounds against such improvised armor, shooting the .223 against several inches of newspaper inside an army field jacket. This target was then placed in front of 10-percent ballistic gelatin (see photographs on pages 146 and 147). As can be seen, the round still penetrated deep enough to provide an incapacitating or lethal wound, while offering no danger of overpenetration.

Banks

A bank is a good example of a hard environment. Every surface, again, seems designed to assist a stray round in ricocheting around until it loses energy. The dangerous situation in banks is, of course, compounded by the presence of innocent bystanders and bank personnel. Going into a bank where I see armed security personnel or police officers has always made a shiver run up my spine. I know that should an incident take place while I am inside, the resulting exchange of fire between felon and guard will have rounds spinning through the air like so many armored bees.

Courthouses

A courthouse is another example, compounded by the fact that such buildings are almost always guarded by armed security personnel and are often quite crowded. A round skipped off the floor in the courtroom would have to be operating by lucky chance indeed not to come to rest in the body of an innocent person.

Airports

And of course we have airports, often the target of terrorists, criminals, and just plain crazies. Travel through any airport in Europe and see what the airport security personnel are armed with. That's right—selective-fire submachine guns. In Vienna's Schwechat Airport attack back in the 1970s, most of the casualties were found to have been caused by bullets fired from the weapons of airport security personnel. Many, but not all, shots were ricochets.

Ships

Shortly after the Italian cruise ship *Achille Lauro* was hijacked off he coast of Egypt by Palestinian terrorists in October 1985, a group of us was given the task of studying takeback of cruise ships. It became apparent that there was a host of problems, starting with how counterterrorist troops were to get on board a ship without alerting the terrorists, who could then be expected to start killing hostages.

A worse problem was in determining what bullets we were going to use. Of necessity, we would have had to use a large assault group, the ships being huge (even more so now) and containing so many compartments. Fratricide by friendly rounds was a distinct danger, necessitating complicated command and control procedures, which would have slowed the assault to an unacceptable degree. Further, rounds ricocheting off shipboard structures could kill or injure many crew members and passengers.

At that point we really never came up with the perfect solution, depending upon the marksmanship skills of our assault teams to cut back on the possibility of missed rounds. Although I believe the troops doing the job at the time (and very probably the ones on active duty now) were up to the task, it would have been far better to have been armed with frangible rounds. Now this is possible.

Of course, hard environments are not confined to the examples above. Look around you. How many surfaces do you see in the course of your ordinary patrol that seem designed to cause a round to ricochet? Unless you are a fish and wildlife officer, probably most of them.

"Sensitive" Environments

To the danger of hard environments, we need to add those where a missed round can do far more damage than merely ricochet or splash back and injure an individual—where it can penetrate or break something that can then cause even more damage or that is irreplaceable.

Nuclear Facilities

Nuclear facilities are perfect examples. How would you like to be the officer whose stray shot made Chernobyl look like a Sunday picnic?

Refineries

Refineries are in some ways even worse. Yes, the potential for long-term damage is greater in the nuclear facility, but at least the structures are hardened and armored. Not so in most refineries. Pipes carrying volatile liquids and gases run all over the place; huge containers are filled with substances that, should they be released, will almost certainly poison the area, at best, or, at worst, produce a conflagration that will destroy the complex and surrounding area. Now add to these natural gas production and transportation facilities, and you start to see the breadth of the problem.

Chemical Laboratories, Production Facilities, and Chemical Storage Sites

Few people are aware of the great number of such facilities even within relatively small communities. And here we are not necessarily talking about military-type chemical facilities producing war chemicals (those being outlawed during the Nixon administration). The insecticide you may be using on your lawn? Mixed with appropriate other chemicals, it becomes a nerve gas. Local jurisdictions "dodge the bullet" time and time again when forces of nature miss local chemical storage warehouses. If the May 7, 1999, Force-5 tornado that carved a swath of destruction through Oklahoma City had taken a path only slightly farther south and west, it would have directly smashed into an area filled with petrochemical storage, farm chemical warehouses, and oil-field-chemical holding facilities. The potential toxic stew produced would have dwarfed the effects of Bhopal.

Would you want to be shooting standard ammunition at a group of terrorists holed up in such a place?

The list could go on and on: hospitals, biochemical research houses, museums. We are told that an accidental discharge in a Smithsonian Institution museum came to rest in something extremely valuable and virtually irreplaceable. No one will admit to exactly what it was. Skip a round off the marble floor of a museum and have that round carve through one of the Old Masters, and see what your job longevity looks like.

Look around your jurisdiction. What you see will probably surprise you. You'll find sensitive facilities and installations in places you would never expect. Now imagine what would happen if critical structures were impacted by stray bullets. You may find that frangible is, indeed, a necessity.

Crowded Situations

The hard and sensitive locations concern mostly the impact of a stray bullet on material and the attendant potential for disaster. There is an even more critical situation, one that you will have no matter what size your jurisdiction: crowds. There may be groups of people at a county fair, or there may be streets filled with pedestrians every day. Whatever the population of your jurisdiction, you can be sure that at some point you are going to have large gatherings of people.

Mass Transit

Consider the subway system in a city. Gigantic tubes of concrete running for miles, interspersed with stone-and-concrete stations. Conveyances are of metal, glass, and plastic that are easily shot through, on the one hand, and ready-made for ricochets if the angle of incidence is even slightly more or less than 90 degrees. Now pack this system with tens of thousands of people. Can you think of a place more prone to disaster? The wonder is not that some people are struck by ricochets and stray bullets during lethal-force encounters in the subway system, but that more are not.

To a smaller degree the same problem is encountered in buses, trains, and airplanes.

On buses, trains, and other mass-transit conveyances, frangible can be a good choice. Once again, unless a stray round directly hits someone it is not supposed to, it is far less likely to cause collateral damage than a conventional round.

As for airplanes, we are often asked if frangible ammunition is suitable for use on airplanes in flight, since firing conventional ammunition that will certainly penetrate the skin of the aircraft poses the danger of explosive decompression. The answer, unfortunately, is no. Examine, if you have the capability to do so, the aluminum skin of an aircraft. You may be surprised at just how thin it is. Any round that can inflict a stopping wound in its intended target contains enough energy to punch through the skin of the aircraft. A round that would not do this can be downloaded. But then it wouldn't do much damage to its intended target, either.

Crowd Scenes

From the county fair to the Super Bowl, law enforcement officers are confronted with masses of people that are difficult to control, at best. We have not yet been plagued with the hooliganism that seems to characterize some European sporting events, but it is probably only a matter of time. Once again we have a situation, as in mass transit, where most surfaces likely to be encountered by a stray bullet seem ready-made for causing ricochet and where numbers of people are standing around just waiting to absorb a bullet's impact.

The panic that such a situation could cause probably would kill more people than any stray rounds. Should, heaven forbid, an officer be forced to resort to lethal force in this environment, he must do it quickly, efficiently, and with absolutely no collateral damage. Having a ricochet whine by your head, even if it misses you by several feet, is enough to get even combat veterans running and looking for cover. Multiply that urge by several thousand people and you can understand why dozens of people are often trampled to death during panic flight.

While talking about crowd scenes, I have to include riots and demonstrations. We must recognize that there are elements who may take advantage of such situations for their own purposes, which can include killing or wounding police officers, firemen, or ordinary citizens. One hopes that the marksmen an agency has placed at strategic locations around the scene (your agency does have such contingency plans, does it not?) are able to surgically remove such threats.

But they may not. A marksman might not have a clear shot, possibly cannot identify the threat, or may not see it at all. Then it becomes the job of the street officer on the scene to take care of the problem. Once again, it must be done quickly and with a minimum of collateral damage. It is a situation where frangible just may be a good choice.

Given what you have seen so far, you should be able to identify on your own those situations and environments in your jurisdiction where frangible is of use. Although I would never advocate frangible only, I do say that a judicious analysis of your own situation could indicate whether or not you would want to include this ammunition in your bag of tricks.

CHAPTER 15

Special Situations/Uses

In addition to the environments discussed in the preceding chapter, in which frangible is a good choice, there are certain specific situations and uses for which frangible is well adapted. The elimination of ricochet and splashback and the lessened likelihood of overpenetration make it a suitable round for the following uses.

DEFEATING BODY ARMOR

This was mentioned in passing in an earlier chapter but bears more amplification here. Soft body armor is readily available to anyone with the money, and the armor's advantages are certainly not lost on felons. High-risk warrant serves and dynamic entries are difficult enough without the added complication of unfriendly people inside wearing body armor. The police officer's usual weapon—the pistol—is not very effective against even relatively light body armor, particularly if the felon is smart enough to put padding between his chest and the armor using some sort of shock-absorbing material (newspaper suffices).

Where do felons get the knowledge of such tactics? It is readily available in the popular press, and, more important, every correctional institution in the world has at least one "school for felons" run by the inmates. Just as you police officers have your academies, criminals have theirs.

By now, everyone in the world has seen the video of the 28 February 1997 North Hollywood shootout, where two would-be bank robbers covered themselves with soft body armor and went about their business, fairly secure in the knowledge that the pistol rounds that most police officers use would not harm them.

Police administrators are hesitant to arm their officers with weapons that are effective against soft body armor, namely, rifles. It is a sad fact, but true, that much misinformation about the effect of rifle rounds is spread among police officers and administrators; that a .223-caliber round will shoot right through the armor-clad felon, exiting with enough force to kill Aunt Millie a half-mile away; that the average rifle round will penetrate one side and exit the other side of almost any building. My favorite is the legend that "you could shoot right through the engine block with that one."

Those who have taken the trouble to test the average rifle round against commonly encountered targets and barriers know better. The .223 in particular (unless you are using green-tip armor piercing rounds, and why should you?) has less of a tendency to overpenetrate than do most pistol rounds. Repeated tests against ballistic gelatin at combat ranges show that the .223 usually folds over at the tip, squirting the lead filler out the back and tumbling. Seldom will it penetrate more than 14 inches, and when it does it has little force.

As for barriers, combat soldiers can tell you many stories about shooting at the enemy, only to have the small, high-speed round deflected by branches and leaves. Yes, you can shoot right through a concrete-block wall with the .223 but only if you hit it repeatedly.

In one study undertaken shortly after the 23 October 1983 bombing of the U.S. Marines headquarters building in Beirut, Lebanon, in which 241 U.S. servicemen were killed, our objective was to determine whether the guards could have stopped the bomber by shooting through the door of the truck. The guards were armed with M16s (although they were not allowed to carry the weapon loaded), so we first tested the .223. What we found was that, if the round hit *only* the relatively thin sheet-metal on both sides of the door, it possessed enough energy to incapacitate a person in the driver's seat. If, however, it hit any mechanisms or struts inside the door, it was either stopped or deflected so much that an incapacitating hit was problematic, at best.

We were encouraged to then try the 7.62mm NATO, there being a lot of old soldiers out there who still argue that we made a mistake going to the 5.56mm round in the first place. And lo and behold, we didn't get markedly better results with the heavier round.

We then tried deflection tests, shooting the rounds against brush barriers that simulated encounters in heavy foliage. The 5.56mm was deflected, as expected. But then so was the 7.62 mm. And, wonder of wonders, so was the .50-caliber BMG!

Our conclusion at the end of extensive tests was that the only round that offered any assurance of hitting the target, punching through the metal of a door and not being deflected, was the 12-gauge shotgun slug. Now there is a round that should give police administrators nightmares: a 12-gauge slug skipping awesome distances, but one that is often allowed because, after all, it is a *shotgun* round and we don't worry about shotguns.

What all of this comes down to is that there is going to be a continuing worry among those who make the decisions about overpenetration and collateral damage when using a rifle round in service, and yet the rifle round is the only means of reliably punching through soft body armor.

This is where frangible comes in. The .223 will reliably punch through Level-IIIA body armor and retain enough force to cause an incapacitating wound in the media behind. It will not, however, penetrate even as far as the back panel of the armor. The standard .223 will give about the same effect, but frangible does not suffer from the same stigma of overpenetration. It is thus easier to convince decision makers that it is a viable round in confronting armor-wearing felons.

ENTRY INTO CONGESTED AREAS

Lack of overpenetration and no ricochet also make the frangible round suitable for use in areas with a lot of potential for collateral damage. I have already discussed crowd scenes and the particular dangers they pose. Again, the police administrator may feel a lot better about the use of frangible in those areas where crowds have to be worried about. Obviously, the officer still has to be sure about the target and beyond and must practice precision marksmanship. A missed frangible round will do just as much damage to a target hit directly as will a standard round, perhaps more so. But if the missed round does not directly strike an unintended target, it is far more likely to break up and become harmless than the service round.

MILITARY OPERATIONS IN URBAN TERRAIN

Most of the emphasis thus far has been on police officers, but the military has justifiable uses for frangible as well. After all, what is an operation in urban terrain if not the same thing that the police do, albeit on a somewhat grander scale?

Probably the biggest problem in MOUT is in command and control. Consider the situation of a military commander who has to clear a building or, worse, multiple buildings. He seldom has the luxury of clearing one room at a time—far too slow. Thus he must have multiple entry and clearing teams going through a structure, separation of the teams often being represented only by relatively thin building materials. Shoot through and attendant fratricide is a real problem.

Frangible rounds, if they do not eliminate the problem, at least vastly ameliorate it. The frangible 5.56mm, upon encountering normal building materials, starts to break up. It will continue to break up the more materials it hits (e.g., the other side of a Sheetrock wall, studs, plaster). By the time it enters the next room, the fragments are so small and have lost so much of their energy that a person clad in PASGT armor will not suffer from any penetration of the armor. This does not allow for the wild spraying of bullets within a room, but it vastly cuts down on the possibility of fratricide. Now the commander can put his troops on line, leapfrog them, bypass pockets of enemy resistance for later elimination, or any of the other tried and true infantry tactics that have always served us well.

CLOSE-IN PRECISION SHOOTING

Recently, a shooting incident got great play on television news. A man was threatening suicide with a pistol. When police officers tried to intervene, he threatened them with it, perhaps hoping for a "suicide by cop" solution. A sharpshooter, armed with frangible ammunition, took advantage of a moment when the would-be suicide was not pointing the weapon at himself or someone else, and fired a shot that ended the incident. The round hit the pistol, knocking it out of the perpetrator's hand and presumably disabling the weapon.

I'm a great believer in the adage that says, "If it works, then it was the right thing to do." More often, however, the close-in precision shots a marksman must take involve something not quite as media friendly as shooting the weapon out of someone's hand. Ending a hostage incident, for instance, with one well-placed shot to the central nervous system will not play as well on the news, but it will resolve the incident.

Would the marksman in the preceding incident have been as willing to do the shot if he had not been able to use frangible? Possibly, but it must be admitted that having frangible in the arsenal makes such decisions easier. At least he knew that if he missed, the round was not going to ricochet off the pavement and possibly strike something it was not supposed to.

If frangible is going to be added to the tools available to the marksman, however, it must be trained with well beforehand. By its very physical makeup, it is not going to act in exactly the same way as the match-grade ammunition with which marksmen are usually equipped. Detailed range data, allowing for differences in wind, humidity, temperature, angle, and all the other factors that can affect the strike of a bullet, should be prepared.

BREACHING

Frangible rounds, particularly 12-gauge slugs, are finding their way into the tools available for the dynamic entry team. "Avon" rounds have been used for this purpose for years (the breacher likes the fact that a well-placed shotgun blast opens a door far faster and with a greater element of surprise than the battering rams and pry bars being used).

For those not familiar with this technique, it goes as follows: The breacher aims the shotgun at whatever secures the door—generally the locking mechanism—and blasts it out with the breaching round. A variation—since quite often the entry team does not know just how many locking mechanisms the door is equipped with (e.g., dead bolts, bars, chains)—is to locate the hinges and blow them away from the door frame. The breacher swings out of the way, and the team makes its entry.

There are a few problems with this technique, all of which can be overcome but which can create issues a team may not want to deal with. The first is that the standard breaching round is made of material that starts coming apart the minute it leaves the shotgun barrel. This is necessary, because shooting a standard slug through a lock or hinge would blast right through the target and go on inside the room where it would stand the excellent chance of hitting something it was not supposed to. This disintegrating characteristic means that the muzzle of the shotgun must be very close to the target—too far away and the effect rapidly dissipates—but not in contact with it. Otherwise the overpressure inside the barrel will most likely burst it and, in a worst-case scenario, injure the shooter and anyone standing nearby. For all practical purposes, this means that the breaching shotgun must be fitted with a standoff device, vented to relieve pressure.

It also means, and this is a worse problem, that the breacher, because of the angle at which he must engage the door, is going to be standing right in front of it. Not the place to be if someone inside decides to react by shooting through the door. Police officers have been wounded in just this way.

It also means that the breacher is for a short period tactically useless. Once the door is breached, he must get out of the way of the entry team, generally slinging the shotgun and drawing a pistol before he can be of any help to the others. What happens if he breaches the door and is immediately confronted by an armed suspect? Attempting to shoot the suspect with the breaching round at anything farther away than muzzle-contact range won't do much good. The breaching round becomes a powder shortly after leaving the muzzle and loses its energy far too quickly to be useful as an antipersonnel weapon.

Frangible helps to solve some of these problems. The round, since it remains whole upon leaving the barrel, does not require that the breacher place the muzzle of the weapon within inches of the target. During breaching exercises at our facility, we instruct the breacher to stand to one side of the doorway, well out of the fatal funnel, and aim at the hinges. Three quick shots and the door is down. No modification to the shotgun is necessary. As long as it hits metal (hinge or lock), the round disintegrates, showing no damage to the witness panels placed inside the door.

Better yet, if the breacher must use the shotgun for protection, he has a viable antipersonnel round still in the tube. The slug, weighing 400 grains and traveling at 1,350 fps, will definitely get the attention of the target, even if he is wearing body armor. Rounds fired at a ballistic panel in front of a Duxseal target show a 2- to 3-inch dent in the soft media of the Duxseal. We can only imagine what effect such an impact would have in the center of the chest cavity.

We've also tested other frangible rounds for improvised breaching and have found that they work quite well. An officer armed with a service pistol and frangible ammunition could very well effect a breach, without the danger of harming anyone inside, by firing at the point where the bolt of the lock enters the door frame. I suggest that you run your own tests on this characteristic and see what works best for you.

Finally, some agencies have decided that the frangible round is just too dangerous to use as a breaching round because if metal or other hard surfaces are not struck by the slug it will likely penetrate the door essentially whole, increasing the potential for injury inside the room. This is true. The breacher using such a round must be experienced and rattleproof, and must ensure that he is able to hit the target every time.

Although I respect this decision, I ask a couple of questions. How much are you paying for your breaching rounds? I suspect far more than frangible shotgun rounds cost, often several times more. Do your officers get to train with the number of rounds they'd like? I suspect not, again because of cost. If your breaching rounds cost $3 to $7 each, you're going to be reluctant to shoot a lot of them.

Frangible shotgun rounds cost, on the average, 75 cents to a dollar each. How many more of them could you use in training for the same cost? Use your high-dollar breaching rounds for the actual breach, if you must. Train, using exactly the same tactics, with frangible.

DISRUPTORS

Finally, one more specialty use of frangible ammunition, though I suspect other uses will be found once frangible is in wider use. Some explosive ordnance disposal technicians have been experimenting with the 12-gauge frangible slug in their "water cannon" disruptors. It has proved a viable option. The slug possesses enough energy to knock the end off a pipe bomb, scattering its contents harmlessly before the device can go off. It can penetrate light barriers to interrupt the firing circuit, and it still disintegrates, vastly lessening the chances of collateral damage.

There are even experiments under way that use the 7.62mm frangible round as a longer range disruptor; its precision allows shots to vital portions of the improvised explosive device that may be too dangerous to approach by other means.

As time goes on, there will be other situations and other environments in which the frangible round will come to be used. By no means am I trying to say that it will replace a large part of the tools now available to the police or military agency. What I am saying is that its judicious application can help solve some of the problems that you may be facing.

But what if you are not a police officer, a military person, a security professional? Does frangible have a use in the civilian shooter world? Is it even legal to use it?

Answers to these questions and more, coming right up.

CHAPTER 16

Nontoxic Frangible Ammunition for Civilian Use

Is NTF legal for civilian use? Can the average citizen even buy it? The answer to the first question is yes. The answer to the second is also yes, but with difficulty.

There are simply no restrictions on the federal level, and none that we know of at the state or local, that keep civilians from buying and using NTF. We are often asked by police officers if we are going to restrict sales to police and military units only. The officers are sometimes dismayed when we say that we are not. The reason for this dismay is the plethora of false information being spread about NTF—that it is more lethal than regular ammunition, that it can't be traced, that it can't be found by X-raying, that it can somehow be used to bypass airport security scanners.

All false. I dealt with the ballistics/tool mark issue earlier, but it merits repeating here. The frangible bullet, fired into flesh-type media, has very prominent and identifiable groove marks in its surface—making it just as identifiable and with probably more prominent marks than on solid-lead or copper-jacketed ammunition. Yes, it is going to require that forensic ballisticians add examples of it to their databases, but they had to do that anyway when police officers started using it, and most already have. Certainly, we have been cooperating by providing samples to forensics labs at the federal and state level.

Increased lethality has also been dealt with. In its training configuration, it is no more lethal than ball ammunition. In its service configuration, it is perhaps slightly more lethal than most hollowpoints, but certainly not more lethal than standard hunting ammunition—and not even as close to lethal as a shotgun blast. It is not a magic bullet, by any means.

Can the fragments of the bullet be seen by X-ray? Consider this: the frangible ammunition we are familiar with is composed primarily of metal. There is generally only enough plastic binder to hold the metal particles together, most often less than 10 percent of total mass. Why should it make any difference to an X-ray apparatus if the metal it sees is made from a melting and molding process, rather than a plastic injection or compression/sintering process? It is still metal. Yes, it shows up quite clearly in X-rays.

What about the howler of not too long ago that it could bypass airport security scanning? Consider this: the projectile is 90 percent or more metal; the case is metal; the primer is metal. The X-ray baggage scanner is going to show it up loud and clear. The walk-through metal detectors and wand detectors will find it just as quickly as standard ammunition. Anyone loading himself down with NTF and attempting to get by airport (or other secure installation) screening is going to be in for a big surprise.

Ergo, there is no legal or moral reason that civilians

should not have access to NTF, and many good reasons that they should.

As to availability, NTF is admittedly hard to find on store shelves or in gunshops. There is a very good reason for this: simple economics. There has been no great demand for this type of ammunition, and store owners are reluctant to stock something they can't sell. When demand increases, there will be increased availability. We are all businessmen, after all, and cannot afford to cede a growing market to our competitors.

Why should the civilian shooter want NTF? It costs more than the ammunition he is accustomed to. It is new and, at least by the average shooter, untested. What are the liability issues? Will it work in my gun? Can I use it for home defense?

NTF IN SELF-DEFENSE

The same characteristics that make NTF a viable choice for police officers and security personnel operate for the civilian. It is less likely to cause collateral damage, which in this case could be a member of your own family behind the typically flimsy interior walls of your house. A round that overpenetrates is not a good choice for self-defense. Neither is one that ricochets. Is an NTF pistol round better for home defense than my personal favorite, a shotgun loaded with #8 birdshot? Probably not, but a pistol is far less unwieldy and is more often the choice of the homeowner. If you are going to defend yourself and your family with a pistol, you owe it to yourself and them to use a bullet that offers the least danger to those you don't want to shoot.

In the states with right-to-carry laws, lots of citizens are arming themselves daily. As John H. Lott says, in his essential work *More Guns, Less Crime*, this has had the effect of greatly reducing crime in those areas, and, frankly, I think it is a trend that should grow into a movement. What will keep it from happening are cases where citizens participate in bad shootings. A hundred good shoots will get barely a mention in the local paper, but one bad one will be featured in the headlines of the *New York Times, Washington Post,* and many other national papers. It will be discussed and decried on the talk shows, and the talking heads of network television will look appropriately grave when they announce this horrible event.

Therefore it behooves us all to make sure there are no bad shoots—no situations where a citizen under duress fires at a felon intent on doing the citizen or a loved one harm, only to miss and have a shot skip down the street and kill an innocent person.

NTF doesn't absolve you of the responsibility for good gun handling, marksmanship, making sure of the target and beyond, and knowing when the application of lethal force is not only appropriate but demanded. But it can help to cut down on the accidents.

If you are to use NTF, you must be prepared to articulate in front of a jury of your peers just why you chose such a round. The prosecutor will likely bring up all the old canards about it: that it is a "plastic round that does horrible damage" or that it is "meant only to kill." Armed with the knowledge this book provides, your defense attorney should be able not only to demolish such arguments, but show the jury why your choice was the responsible decision.

PRACTICE

If you are to use NTF for self-defense, you must practice with it. And NTF has characteristics that make it far easier to use in practice than standard lead ammunition.

It produces no toxins. If you are a country dweller with sufficient room on your property, this means you can set up your own firing range without worrying about some official from the EPA coming to shut it down. You are not polluting any waterways; the copper in the earthen berm is not going to contaminate groundwater. Rounds fired into the berm are far less likely to skip when they hit something hard, no doubt pleasing neighbors who might otherwise have been in the impact zone.

It even makes it practical for the first time for a citizen to build an indoor range. Our test range is nothing more than a steel box with one side open. Rounds fired into the box disintegrate against the backplate, the sidesplatter is caught by the box wings, and nothing comes back to the shooter. The total cost of this range was the cost of the steel. I recommend armored steel for the backplate (AR 400 if only pistol rounds are to be used, and AR 500 if rifle rounds are also to be shot) in 3/8-inch thickness. The wings, not being subject to direct bullet strikes, can be of mild-steel plate. Put up some noise-absorbing material on the walls, and you have a basement range! Obviously, before embarking on such a venture, you need to check local ordinances.

RANGE VERSATILITY

Firing ranges catering to the citizen shooter are finding out that they can increase their business and afford the opportunity for rifle marksmen to practice on a range designed for pistol shooters. The bullet traps on modern ranges will easily withstand hits from NTF rifle rounds, even including the high-energy .223. Rifle shooters have always been plagued by a lack of facilities for shooting, even more now that urban encroachment and government regulation have closed down most outdoor ranges within easy driving distance of the big cities. NTF remedies this problem.

MATCH USE

Various pistol, rifle, and shotgun matches are staged throughout the United States every year, a trend that seems to be growing. Many of these matches involve the use of tactics as well as marksmanship, which is a trend that we would like to encourage. The more citizens who use and practice with defensive weapons, the fewer misconceptions there are among the public, and the less hue and cry for gun control.

NTF is particularly suited for many of these matches. Not only is it as accurate as (or more accurate than) the rounds now being shot, but it cuts down on the dangers and the environmental problems faced by match sponsors.

Cowboy Action Shooting

Cowboy Action Shooting is the fastest growing segment in the shooting sports. The primary targets in these shoots are of steel, and solid-lead rounds are the projectiles of choice (authenticity, you know). I went to one of these shoots last year, and with the whining and whirring of ricochets for a moment thought I was back in Vietnam during the Tet Offensive of 1968. The safety of these matches could be improved immeasurably by the use of frangible ammunition, and in fact we are now working with Single Action Shooting Society (the primary sponsor of such events) to get the rule books changed to allow its use. Another benefit would be in acquiring locales in which to conduct matches. Most would-be landlords of such events become extremely unenthusiastic when notified that they may be responsible for the cleanup of all the lead expended in the event.

Silhouette Shooting

Silhouette shooters are often wary of frangible ammunition, thinking that it does not possess enough power to knock down the target because it is much lighter than standard ammo. This caution is misplaced: with its increased speed frangible possesses at least as much transferred energy as conventional rounds at the ranges ordinarily shot in matches. To add to this, the trajectory is generally flatter, making it easier to "shoot where the sights are" rather than adding estimated elevation. Since the distances ordinarily shot at these matches are out well past the range where splashback is a problem, frangible offers no real advantage in this area. However, it still cuts down on the environmental problem and, more important, its reduced maximum range allows it to be used in areas far more restricted than is now the case.

Practical "Tactical" Matches

Several organizations have grown up to sponsor tactical-type matches, drawing contestants from police, military, and civilian circles alike. All can benefit from the use of frangible. As discussed in earlier chapters, real-world encounters occur at extremely short ranges, but tactical matches cannot be conducted at such ranges (at least against steel reactive targets) because of the danger of splashback and ricochet. Frangible can help make the match more realistic, more representative of actual shooting situations, And, of course, it can help expand the venue to areas that before were not suited to such matches and increase the draw of the sport itself.

RIFLE USE

Frangible rifle ammunition can be used in places that heretofore were restricted to pistol use. This includes matches, but is also applicable at any local range. Frangible rifle ammunition can be used against bullet traps normally rated for full-power pistol ammunition and offers no more danger of splashback and ricochet than do frangible pistol rounds. We quite commonly shoot frangible .223 at steel plates from ranges of 6 feet or less, and out of thousands of rounds have never had a problem.

CAN I USE IT IN MY WEAPON?

This is the sticking point that causes many manufacturers to restrict their sales to law enforcement and military only. Frangible is a modern, high-powered ammunition. Its chamber pressures are just as high as, and sometimes higher than, the round you may be accustomed to. Its pressures are, of course, kept within SAAMI maximums, but SAAMI standards are based on modern weapons in good condition.

Police and military agencies will almost always be armed with relatively new, well-maintained weapons. A manufacturer can sell to them with relative confidence that its ammunition will not be used in anything else. In the civilian world, no such assurances are available.

So we have the situation of Joe Public buying a box of frangible, taking it home, and pulling out the P08 Lüger Grandpa brought back from World War I. Said weapon has been quietly rusting away for the past 80-odd years. Joe loads up, fires a couple of rounds, and has the chamber blow away, injuring himself and perhaps his wife and kids.

Wanna bet there's going to be a lawsuit? And even if Joe had bought standard ammunition and had done the same thing with similar results, the jury won't pay attention to that. This is a new product. How can we be sure it was suited for that particular weapon? The manufacturer should have known, blah, blah, blah. And Joe gets a multimillion dollar reward for his stupidity.

Let me say it here, loud and clear: *frangible ammunition is a modern round, designed for use only in modern weapons in good condition.* Load it into a cheap knockoff, and ask for trouble. Load it into a weapon designed for pressures far below those of other modern rounds, and expect it. And no older weapon should be fired, using any ammunition at all, until it has been checked by a qualified gunsmith and pronounced safe.

As a shooter, we expect you to be a responsible person. Antigun zealots would have it otherwise. Don't give them more ammunition.

Conclusion

At the beginning of this book I posed this question: "Is frangible ammunition the round for all reasons?" By now, of course, you know that it is not. No existing bullet is any such thing, each having its own advantages and disadvantages.

It is unlikely that we will have a weapon that fits all needs, under all circumstances, until we have reached the *Star Trek* phaser level of technology. Even then, I suspect, there will be those who long for the good old days of projectile weapons, and others who will say that the phaser is good as far as it goes but doesn't go far enough. Just wait for Phaser M2!

What I hope you have come away with is that NTF has a place in your arsenal, in some cases only for training, in others perhaps for service use as well. It can help solve many of the problems with which you are probably confronted. But all too often it is easy to bow to the latest restriction or the newest regulation and downgrade training.

Twenty years ago, when I was preparing to retire from the U.S. Army, I took as my last assignment a job teaching ROTC cadets at a small college in the Southwest. The firing range (restricted to .22-caliber weapons) was on the ground floor of a campus building. Even then, before we were really aware of the lead problem, I knew it was an unhealthy environment. Ventilation was poor to nonexistent. When you came out of the range after a couple of classes, you blew your nose, and it came out black. The lead was so overwhelming that you continually tasted it on the back of your tongue—an unmistakable taste. Yet we were spending hours there each day—the senior NCO even had his office there!

The problem was recognized, and suddenly there was a directive to get rid of all the .22-caliber weapons. However, we still had to teach marksmanship to the cadets. The solution was to procure precision air rifles shooting .17-caliber lead pellets. Yes, we went from shooting jacketed .22 rounds to shooting pure-lead pellets! Amazing, you might say, to substitute one thing for something that could very well be quite worse, but that was the state of knowledge back in those days.

That range has long since been closed down, as have most other ROTC ranges in the United States. I have been told that after one inspection of such a range, the inspectors taped and sealed the door, not even allowing instructors to go back in and secure their personal gear.

As a result, marksmanship training has suffered immensely. Cadets go on active duty having fired nothing more than what can be shot during summer camp. An officer doesn't need to be able to shoot, you say? Agreed, he is supposed to be influencing the action, rather than serving as a rifleman. But the attitude this engenders, and the even greater gulf it creates between leaders and the troops, has and will continue to contribute to the decline of our military.

I bring up this example only to point out that it is a situation shared, not only in the military, but in many police agencies. Range time has become so difficult to schedule, what with the restrictions and the closing of ranges, that it becomes easier to not schedule it at all. Will the results be something like the lawsuits quoted in an earlier chapter? Inevitably. More important, officers and innocent civilians will suffer the dire, and often lethal, consequences.

Anything that we can do, and this includes the reopening of ranges or the construction of new, low-cost facilities, to remedy this problem is the only responsible course. NTF can help.

There are some writers and thinkers who surmise that EPA and OSHA regulations are a smoke screen, that the real purpose behind all these restrictions is to make it far more difficult and costly to shoot, thereby reducing the citizen's motivation to go out and practice—and thus reducing his desire to own a gun. I take no position on this as a deliberate policy, but I do say that it has had that particular effect. The harder it is to find a place to shoot, the less you are going to shoot. The less you shoot, the more danger you are to yourself and to others, and the more likely it is that unfortunate incidents will occur. Such incidents are then seized upon by the antigunners as just more examples of why no one should possess a gun.

Once again, for civilians, as well as for police and military agencies, NTF can be used to ameliorate this problem. Many civilian firing ranges are already using NTF, with more to follow. Sources at the major range-manufacturing concern in the United States tell me that they are being asked more and more often to design ranges specifically geared for NTF. And when you can build a range for 50 percent less than you can build a conventional range, the economics of range construction suddenly become a lot more desirable. We can hope that this leads to more shooting facilities' becoming available rather than the current situation, in which they are closed almost daily.

Besides, if it is true that the authorities are using environmental regulations to further hamper gun ownership, isn't it delicious that we can use technology to stymie them?

What you should also come away with, whether you are police, military, or civilian, is that you should make the decisions on whether to use NTF or not in your own tests. Don't accept manufacturers' claims for the ammunition. You wouldn't do that for anything else you buy, so why should you do it for something your life might depend on?

I have listed the characteristics to look for and indicated the tests you might do, and none of them is long and complicated. In the Appendix, I list sources currently available for testing, training, and obtaining supplies.

Good luck and good shooting.

APPENDIX

Sources

NONTOXIC FRANGIBLE AMMUNITION

Longbow, Inc.
P.O. Box 129
Martha, OK 73556
(580) 482-2600
Fax (580) 482-2700
E-mail: bulletbaron@yahoo.com

Black Hills Ammunition
P.O. Box 3090
Rapid City, SD 57709
(605) 348-5150
Fax (605) 348-9827

Blount International, Inc.
900 Ehlen Drive
Anoka, MN 55303
(800) 627-3640
Fax (612) 323-2506

SNC Technologies, Inc.
5 Montée des Arsenaux, Le Gardeur
Quebec, Canada J5Z 2P4
(450) 581-3080
Fax (450) 581-0931

West Coast Bullet, Inc.
P.O. Box 2270
Dayton, NV 89403
(702) 246-3941
Fax (702) 246-0836

Winchester Ammunition
427 North Shamrock Street
East Alton, IL 62024
(618) 258-2000
Fax (618) 258-3609

FRANGIBLE-ONLY RANGES

WCB Associates, Inc.
P.O. Box 129
Martha, OK 73556
(580) 482-2599 Fax: (580) 482-2700

Caswell Detroit Armor Companies
2540 Second Street Northeast
Minneapolis, MN 55418
(612) 379-2000 Fax: (612) 379-2367

About the Author

John F. Mullins spent 22 years as a special operations noncommissioned officer and officer, retiring from the U.S. Army Special Forces with the rank of major. After retirement, he became a consultant in security and training. At various times, he has worked for the Department of Defense, State Department, Department of Transportation, Federal Aviation Administration, and the Department of Justice.

Mullins started developing his own frangible ammunition in 1989 and in 1997 patented the bullet that is now used by Longbow Incorporated to supply police, military, and selected civilian ranges throughout the world.